PUTTING THE
JUST-IN-TIME
PHILOSOPHY INTO PRACTICE

PUTTING THE JUST-IN-TIME PHILOSOPHY INTO PRACTICE

A STRATEGY FOR PRODUCTION MANAGERS

P J O'GRADY

KOGAN PAGE

To Mary, Catherine and James
for their patience and encouragement.

Copyright © 1988 P.J. O'Grady

All rights reserved. No reproduction, copy or transmission of this publication may be made without written permission.

No paragraph of this publication may be reproduced, copied or transmitted save with written permission or in accordance with the provisions of the Copyright Act 1956 (as amended), or under the terms of any licence permitting limited copying issued by the Copyright Licensing Agency, 7 Ridgmount Street, London WC1E 7AE.

Any person who does any unauthorised act in relation to this publication may be liable to criminal prosecution and civil claims for damages

First published in 1988 by
Kogan Page Ltd,
120 Pentonville Road, London N1 9JN

British Library Cataloguing in Publication Data

O'Grady, P.J.
  Putting the just-in-time philosophy into
  practice : a strategy for production
  managers.
  1. Factory management
  I. Title
  658.5   TS155

ISBN 1 85091 121 5

Photoset in North Wales by
Derek Doyle & Associates, Mold, Clwyd
Printed and bound in Great Britain by
Billing and Son Ltd, Worcester

CONTENTS

PREFACE 9

1 INTRODUCTION 11
   Problems of manufacturing management 12
   Just-in-time 13
   Implementation: the five steps 15

2 TRADITIONAL MANUFACTURING MANAGEMENT 17
   Introduction 17
   Inventory control policies 18
   Materials requirements planning systems 23
   Manufacturing resource planning: MRP II 25
   MRP problems 26

3 JUST-IN-TIME OVERVIEW 32
   Introduction 32
   JIT is not a software package 33
   JIT is not a methodology 35
   Attack fundamental problems 36
   Eliminate waste 38
   Strive for simplicity 42
   Devise systems to identify problems 47
   Cost/benefit of implementing JIT 49
   Conclusion and summary 50

4 STEP 1: GETTING THE BALL ROLLING 52
   Introduction 52
   Basic understanding 53
   Preliminary education 54
   Cost/benefit analysis 56

Commitment 60
Go/no go decision 60
JIT project team selection 62
Identifying the pilot plant 65
Conclusion 66

5 STEP 2: EDUCATION – THE MAKE OR BREAK ISSUE 68

Introduction 68
What is JIT education? 69
Who needs JIT education? 71
What should be covered? 73
Conclusion 75

6 STEP 3: PROCESS IMPROVEMENTS 77

Introduction 77
Set-up time reduction 78
How is low set-up time achieved? 80
Preventative maintenance 82
Changing to flow lines 85
Conclusion 89

7 STEP 4: CONTROL IMPROVEMENTS 91

Introduction 91
Simple control 92
Pull systems 93
Kanban systems 95
Linking MRP with pull/Kanban systems 97
Shop floor control and quality at source 99
Conclusion 101

8 STEP 5: VENDOR/CUSTOMER LINKS 103

Introduction 103
Links with vendors 104
Multi-sourcing *versus* single-sourcing 107
Short- *versus* long-term agreements 108
Local *versus* distant suppliers 110
How to implement links with vendors 110
Links with customers 111
Conclusion 112

9 JIT IMPLEMENTATION – THE PROVEN PATH 114

Introduction 114
How long should the implementation take? 115
The implementation sequence – the proven path 117
Case study A 119
Case study B 122
Conclusion 124

10 SUMMARY AND CONCLUSION 126

Just-in-time systems 126
Potential pitfalls 129
The future 131

BIBLIOGRAPHY AND FUTHER READING 132

INDEX 135

Preface

This book describes both the essential features of Just-In-Time (JIT) approaches to manufacturing and how JIT can be successfully implemented. JIT marks a significant departure from previous western approaches to manufacturing management, and aims to improve quality levels and customer service while decreasing lead times and inventory levels. The use of simple though effective methods can, with proper management, lead to continual improvements in the manufacturing operation.

A number of companies have now implemented JIT and some of these implementations have been very successful. However, what is becoming increasingly clear is that there is a significant number of JIT implementations that fail to achieve the potential benefits of JIT. It is not an easy task, and there are a number of pitfalls that await the unwary manager. My motivation for writing this book has been my experience of working with companies that have been successful in implementing JIT and of seeing what needs to be done and how the most common pitfalls can be avoided.

The book is oriented towards batch manufacturing since this accounts for a large proportion of manufacturing in most western countries. Other types (including process, mass and jobbing) can also profitably use many of the JIT techniques to improve their operation.

This book is aimed at production, operations and manufacturing managers but will benefit those engaged in such support activities as computer systems, finance and sales, as well as being very useful for students in a variety of disciplines. My hope is that this book will aid the understanding of JIT and thus will lead to better JIT implementations.

Peter O'Grady
Department of Industrial Engineering
North Carolina State University
Raleigh, NC 27695–7906
USA

1
INTRODUCTION

Western manufacturing industries are under siege. Many are still reeling from the impact of foreign competition and some have virtually disappeared, perhaps for ever. The industrialized countries of the West rely on manufacturing for their wealth – the wealth which supports the activities of, for example, the stock market, the commodity exchanges or the real estate sector. The fact that manufacturing is in trouble therefore strikes deep into the fabric of our society. With a weakened manufacturing base we risk a deterioration in the economic well-being of the developed countries of Europe and North America and there is a real possibility that the balance of economic power will shift decisively to those countries still able to sustain a strong manufacturing sector.

It is the ability to manage the manufacturing process that is the key. Those countries that have achieved continued success in manufacturing, such as Japan, have done so *not* on the basis of sophisticated technology (although the judicious use of such technology can help) but because they have managed their human resources better than their western counterparts.

If we want western manufacturing industries to survive, our first priority must be to improve the way in which they are managed. The new approaches to manufacturing management that were adopted in the 1970s and early 1980s certainly helped, but if manufacturing industry is to flourish and to regain the market share it has lost, more fundamental changes will be required.

Some have suggested that the solution lies in the use of complex computer systems. These systems would be able to continually monitor production, preplan activities and adjust production to give the desired response. Such an approach is the basis of the manufacturing resources planning (or MRPII as it is usually known) system that has been implemented in many companies. MRPII can certainly bring about improvements; the improvements reported so far, however, are not sufficient to revitalize western industries. [See Anderson *et al.* (1982)

whose survey indicated that MRPII users increased inventory turns from 3.2 to only 4.3 on average.]

Instead, it is increasingly being recognised, the key lies in the application of the Japanese approach to manufacturing management, adapted to take account of the cultural and socio-economic differences between Japan and western countries. This adapted Japanese approach is called *Just in Time* or *JIT*. Already, many companies have made a determined effort to implement JIT and have found it to be a highly successful approach; their experience suggests that JIT could play a major part in revitalizing many of our industries. However, JIT is not particularly straightforward to implement and in some cases it has signally failed to produce the improvements which were anticipated.

This book has two aims. The first is to provide a clear definition of JIT, together with a description of its main characteristics and an analysis of its potential benefits. The second is to provide a step-by-step plan for the successful implementation of JIT. The strategy that underlies this plan has been distilled from close observation of many JIT implementations in a number of western countries, some of which have been wholly successful whilst others have been only partially so. A careful examination of the successful projects led to the construction of a five-step implementation plan that has subsequently been successfully put into effect in many companies. This book examines each of these five steps and describes how they should be coordinated in order to achieve the best possible implementation.

Problems of manufacturing management

The term manufacturing industry covers a broad spectrum of activities, ranging from the control of chemical processes to precision engineering, from making satellites to making bicycles [see Vollman *et al.* (1984)]. Overall we can divide the manufacturing process into three major categories:

- Flow (or mass) manufacture
- Jobbing manufacture
- Batch manufacture.

Flow manufacture consists of high-volume manufacture of a small range of products such as, for example, consumer goods like television sets or vacuum cleaners. Flow manufacture has certain very obvious characteristics: because it is concerned with achieving high outputs of a limited product range, it often makes sense to invest in specialized machinery, and since there is a significant degree of repetition involved

Introduction 13

a high proportion of the labour force can consist of unskilled or semiskilled workers.

The management problems in flow manufacture are therefore mainly concerned with ensuring a continuity in material and component supply as well as a high overall system efficiency. It is worth noting, however, that in spite of the high output achieved by many flow manufactuers in the West, poor quality has been a continuing and intractable problem.

Jobbing manufacture is concerned with manufacturing non-standard items in a one-off mode; that is, the manufacturer cannot assume that he will receive repeat orders for products. Jobbing manufacture is therefore characterized by a large range of products, each product being low volume, and the requirements in terms of the machinery used and the skill levels of the workforce are thus quite different to those found in flow manufacture. General- rather than special-purpose machinery is needed and workers in this non-repetitive environment must be highly skilled. Again, major management problems are ensuring high system efficiency commensurate with high quality levels.

The final category of manufacturing, batch manufacture, is concerned with medium-volume products where repeat orders *are* expected. In such batch environments, a product is manufactured in a batch, or lot, at intervals which vary according to demand and other factors. A batch may be repeated every day, week, month, year or every few years. This kind of manufacturing poses considerable problems for management since high overall system efficiency must be maintained in the face of constantly changing demand patterns.

In practice, many companies are engaged in manufacturing activities that spread over all three categories; but, overall, batch manufacture accounts for between 75 and 85 per cent of output in western countries. But, whichever category is involved, there is one feature that characterises nearly all manufacturing enterprises – complexity. Manufacturing systems are inherently complex, and a huge range of factors determines the interrelationships between jobs, machines and personnel. As a result of this complexity, manufacturing management in western countries finds itself desperately in need of techniques that will improve inventory turnover, quality control and general efficiency. JIT can meet that need.

Just-in-time

In fact, JIT offers an opportunity for manufacturing management to take a completely new direction. For JIT is *not* simply a methodology nor is it a piece of software. One cannot, for example, describe JIT in terms of the logical relationships that can be used to define the operation of materials requirements planning systems, nor can JIT be purchased as a software package which only requires the supply of accurate data for effective operation.

Rightly seen, JIT is a *philosophy* that defines the manner in which a manufacturing system should be managed. This means that JIT is both more far-reaching and more nebulous than other systems such as materials requirements planning; and the implementation of JIT is therefore likely to involve correspondingly greater difficulties. In particular, more time must be spent on such areas as education than is the case with other approaches. There have, however, now been many successful JIT implementations from which we can learn.

The essential objectives of JIT are fourfold:

- attack fundamental problems
- eliminate waste
- strive for simplicity
- devise systems to identify problems.

The first objective can be described as basic good management in that, rather than masking problems, JIT attacks their fundamental causes. Where there is a chronic capacity bottleneck, for example, there is little point in attempting to obtain better schedules in order to overcome the problem. Instead JIT, would indicate that the only way of solving a capacity bottleneck is to increase the capacity, either by the use of additional machines or personnel or by subcontracting the work to another facility.

The second objective, the elimination of wasteful activity, is likewise no more than applied common sense. Examples of activities which JIT aims to eliminate, or at least reduce to a minimum, are inspection, transport and inventory and the manner in which these activities can be minimized are described in Chapters 6, 7 and 8.

The third major theme of JIT, striving for simplicity, emphasises the need to simplify the operation of the manufacturing system by, for example, reorganising the complex ebbs and flows of parts and products through a factory into simple unidirectional flows. The examples given in later chapters show how such changes can reduce and simplify management problems.

Before fundamental problems can be solved, they must first be identified and this is the final objective of JIT: devise systems to identify problems. Successful JIT implementations are characterized by mechanisms that identify fundamental problems which are then brought to the attention of the management. Thus bringing us back to the first principle of JIT, that of solving fundamental problems.

All four major objectives can be achieved without incurring major costs. Since JIT stresses simplicity, little capital investment is usually involved although, as stressed in later chapters, some additional equipment may have to be installed and some expenditure is likely in reducing setup time. As will be detailed in this book, the total costs involved in a successful implementation are often low compared with the returns so that JIT can be described as a low cost/high return policy.

Implementation: the five steps

The achievement of a good rate of return is, however, dependent on a good implementation. This book describes five steps that are crucial to this:

- Step 1: Getting the ball rolling
- Step 2: Education
- Step 3: Achieving process improvements
- Step 4: Achieving control improvements
- Step 5: Extending vendor/customer links

The first step involves constructing the foundation upon which the implementation can be built, a process which is described in Chapter 4. Since implementing JIT involves changing attitudes within a company this first step sets the whole tone of the implementation. It includes some initial education, the cost benefit analysis, and the identification of the pilot plant. But perhaps the most important factor in getting the ball rolling is the securing of top management commitment. Without this, the implementation may be considerably more difficult, since some hard choices will inevitably have to be made at key points. Strategies for obtaining top management commitment are also detailed in Chapter 4.

Once the first step has been successfully completed the task of education can commence. The fact that this step has been called 'the make or break issue' indicates its importance. A successful JIT implementation requires the changing of attitudes which have often become deeply ingrained and this can only be achieved through the kind of comprehensive education programme described in Chapter 5.

Once the education programme is underway then changes can be made to the processes (Chapter 6) – and to the manufacturing control (Chapter 7). These improvements include the use of mini-factories with flowlines to simplify the control problems as well as the use of Kanban/pull systems to 'pull' work through the manufacturing system.

The final step of extending vendor/customer links completes the JIT implementation. This step, discussed in Chapter 8, incorporates the vendors and the customers in a JIT system that covers the whole manufacturing process from vendors, through the company itself and on to the customers.

These five steps form the basis for JIT implementation. They have been tested in practice and form the core of the implementation plan described in Chapter 9.

2
TRADITIONAL MANUFACTURING MANAGEMENT

Introduction

The management of most manufacturing operations poses tremendously complex problems. Nowhere is this more true than in batch manufacturing. At any one time, a typical batch manufacturer may have several hundred batches flowing through dozens of work centres. Coordinating personnel and machines to get the right product to the right customer at the right time, in the right quantity while also achieving acceptable standards of manufacture at reasonable cost is a mammoth task.

Over the years, managers have adopted a variety of approaches in the hope of resolving, or at least simplifying, their problems. The earliest attempts were based on simply monitoring the inventories of the finished products: when the inventory level of a particular product became too low, a new batch was ordered.

Such approaches were (and still are) very simple to operate; but they can only provide a partial solution. In the 1970s, when both offshore competition and interest rates increased, companies came under even greater pressure to reduce inventories. As a result attention began to be focussed on materials requirements planning (MRP) and manufacturing resource planning (MRPII) type systems [Orlicky (1975)]. These MRP systems produce a detailed plan for material and component requirements.

This chapter gives an overview of these well-established approaches to manufacturing management. But it should be borne in mind that they have not always succeeded in significantly improving manufacturing efficiency [Anderson et al. (1982)]. Indeed, the available data indicate that western countries still lag behind the countries in Southeast Asia in such performance measures as stockturn, customer service and quality levels. A proper understanding of these 'traditional' approaches is, however, necessary in order fully to appreciate the JIT approach, if only because it enables us to identify some of the pitfalls that must be avoided in the course of a JIT implementation.

Inventory control policies

The earliest mechanisms used to manage manufacturing did not involve any analysis of the manufacturing aspect at all, but instead concentrated on monitoring the inventories of finished products. When these inventories fell below certain levels (usually called reorder levels or reorder points) a replenishment was ordered and this order passed through the factory and then into the inventory, hopefully before the inventory was exhausted. [See Vollman *et al.* (1984)]. A manufacturer producing several hundred different lines could therefore make all his management decisions on the basis of information about his finished product inventory levels (see Fig. 2.1).

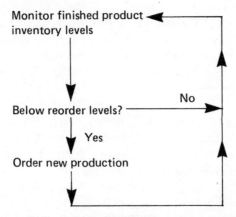

Figure 2.1 *Inventory monitoring and reordering*

The frequency with which the inventory was monitored varied between companies. One typical system involved maintaining a running total of the inventory levels; as soon as the inventory of a particular item fell below four months' expected sales then production of that item was ordered – four months being the time that had to be allowed for a new batch to be manufactured.

Another system, in use with a US company that subsequently switched to a JIT approach, required a physical count of all the finished items in stock every week (a time consuming and relatively expensive chore); management then decided, on a case-by-case basis, which products were likely to have their inventories depleted in the near future. These products were then ordered for production. This particular company kept no records of actual manufacturing or supplier lead times so their estimates about expected inventory depletion often turned out to be inaccurate. This was especially significant for one of their product ranges that had abnormally long lead times (about six months, or twice the time that was required for the other products). But, since the company kept no adequate records, the management did not realize that extra lead times were needed for these products and consequently the stock of this product range was often depleted before the ordered replacements arrived. <u>This naturally led to poor service and bad relations between the company and its customers</u>. In fact, the company's recent history had been one of bad sales performance overall, and it took the implementation of a JIT system to turn the business around.

This inventory control approach to manufacturing management pays little or no regard to *how* production is going to be addressed; it also frequently results in significant fluctuations in production requirements. If, as often happens, inventory levels of a large number of products simultaneously fall below the critical level, then large amounts of production are ordered, creating production bottleneck problems. Under these circumstances manufacturing lead times may vary widely, depending on capacity bottlenecks and other factors. This question of fluctuating lead times is one we shall return to several times in later chapters, for wide variations in lead times make it extremely difficult to organize production efficiently and to replenish inventories promptly.

When long lead times are involved, other problems are compounded. For example, it becomes necessary to forecast customer demand over a longer time span; and while it may be relatively easy to forecast demand for next week or even for next month, it is a great deal harder, and far more risky, to forecast demand four or six months ahead. Over such time spans significant and often unforeseeable changes take place – in interest rates, in the level of economic activity, in raw material and oil prices, in taxation levels or in dollar/yen exchange rates. All or any of these can fluctuate dramatically over a four-month period, often making nonsense of forecasts that may have been arrived at after long and exhaustive deliberations. The point to bear in mind is that if we significantly <u>reduce manufacturing lead times</u> then the problem of forecasting future demand can be significantly reduced.

Another problem caused by long lead time is financial: if products take four months to move through the production process then the capital invested in those products is locked up for four months. If lead times can be reduced to, perhaps, one month then the amount of capital tied up in work in progress is reduced by 75%. For many industries this is a significant factor. If, for example, a manufacturer has $300,000 of work in progress then reducing this by 75% will release $225,000. This money can then be put to use, perhaps to purchase new machines and equipment that will help to further enhance the productive capability.

Reducing manufacturing lead times also allows stocks of finished goods to be cut since the uncertainty has been reduced and with it the need to maintain additional inventory against an unexpected surge in demand.

Another benefit of cutting back lead times is that it reduces space requirements. Less work in progress reduces the clutter and allows the managers of shop floor activity to identify problem areas more easily and to keep better track of jobs as they progress through the factory. As a result, the company's activities are made more effective and efficient. Japanese factories are thought to have only one third the floor area of equivalent western factories, resulting in lower maintenance, heating and transport costs within the factory.

One further major advantage from reducing manufacturing lead time is that it is likely to lead to quality improvements, especially where raw materials or work in progress deteriorate with age.

Long manufacturing lead times can also lead to a decline in the effectiveness of management. If the forecasts of future demand turn out to be inaccurate (as is usually the case) the orderly flow of work through the factory is disrupted. Products which are no longer needed because the forecast over-estimated demand are put to one side to await an upturn in demand, these stock-piles of partially completed goods disrupt the flow of work through the factory (as well as locking up valuable capital). Conversely, part finished products for which the forecast has underestimated now become in demand and have to be pushed through the factory as a matter of urgency. Such sudden changes in priority tend to create confusion and are likely to lead, first, to quality problems and, second, to inefficient use of resources when partly processed items are taken off a machine and replaced by those that are currently required. The entire plant becomes pervaded by an atmosphere of crisis, and managers become totally pre-occupied with the need to resolve short-term problems.

Figure 2.2 is a diagrammatic representation of the way in which the inventory control approach can be elaborated to take account of the

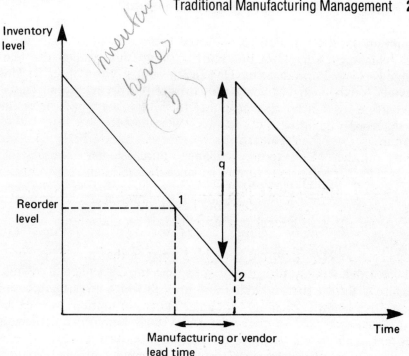

Figure 2.2 *Operation of reorder point approach*

lead-time that must be allowed for the manufacture of a new batch of a specified product – this, of course, assumes that the management has accurately established the lead-times for each product in the company's range.

The steps taken are as follows:

- When the inventory level of a given item falls below the reorder level the order to produce a new batch is sent out.
- After the manufacturing or vendor's lead time has elapsed, the batch arrives in finished goods inventory and boosts the inventory level. The process repeats itself in subsequent time periods.

In order to determine the quantity, q, that should be ordered it is necessary to achieve a balance between the storage and manufacturing costs. For example, if q is large then the set-up and other fixed manufacturing costs per item will be low. However, this large quantity will eventually arrive in finished goods inventory and boost the level considerably, leading to higher inventory holding costs. If, on the other hand q is small then inventory holding costs will be reduced but the set-up and other ordering costs per item will rise.

Several analytical methods have been developed to determine the 'optimum' level for q. Perhaps the most commonly used technique is the Economic Order Quantity Formula (sometimes called the Economic Lot-Size Formula or the Harris-Wilson Lot-Size Formula). This formula is based on balancing the costs of reordering (set-up costs) against the costs of holding inventory to produce an 'optimum' value for the reorder quantity, $q^\star$, as follows:

$q^\star = \sqrt{(2rc_3/c_1)}$
where
$q^\star$ = 'optimum' reorder quantity
r = average demand (expressed in items/year, month or week)
c_3 = cost of set-up and reordering
c_1 = inventory holding cost (expressed in cost/item/year, month or week)

I have deliberately placed the word optimum in inverted commas because of the number of hidden assumptions which underlie this, and many other, reorder quantity formulae. These assumptions include the notion that each product can be considered in isolation, that demand is constant, and that lead times do not vary under different circumstances. In a typical manufacturing environment where products compete with each other for valuable production resources, where demand can vary dramatically from one month to the next and where the lead times for one product may depend upon the plant's workload and the range of items currently in production, the Economic Order Quantity type formulae provide only a partial solution [see Vollman *et al.* (1984)]. Certainly, the solutions obtained by such a formula must always be treated with care and adjusted when necessary. Unfortunately, however, the superficial appearance of precision which attaches to any mathematical formula has given many managers the impression that when they use these formulae they are obtaining the best possible performance.

One advantage which this approach does have is its simplicity. The inventory of each item is managed separately using very straightforward mechanisms. In practice, the responsibility for reordering can often be left with warehouse personnel who are ideally placed to monitor changes in inventory levels and react to them. This approach also has the advantage that it readily lends itself to a manual operation. Indeed, prior to the widespread use of computers in MRP type systems (see below), this inventory control approach to the determination of reorder levels and quantities was the only manufacturing control approach in use in western industries.

The disadvantages of the inventory control approach are however

Traditional Manufacturing Management

severe. The main ones can be summarized under three headings:

1. *Higher inventory holding costs.* Individual inventories of all components are kept with much of this inventory not being used for some time. The inventory that is being held is costly in two major respects; it consumes capital that could be used elsewhere to increase productivity, and it occupies more expensive factory space.
2. *Lack of responsiveness.* In any dynamic market there are likely to be dramatic changes in demand. Where the market demand suddenly increases then it is likely that there will not be enough inventory to cover any increase and the customer service level therefore falls.
3. *Risk of inventory becoming obsolete.* When there is a downturn in demand a manufacturer may well be left with large quantities of surplus stock. In many cases this stock will become obsolete and may be kept in inventory for many years (still tying up money and space) before being written off and disposed of.

Both of the latter two disadvantages arise because the inventory control approach is essentially *reactive*, it responds to changes as and when they occur without anticipating or looking to the future. In the increasingly competitive atmosphere of the 1970s, when large inventories became very expensive to maintain, it became clear that there was a need to move towards a more *proactive* approach. This would allow changes to be anticipated and the manufacturing process to be managed in order to ensure that the response was fast and accurate. The proactive systems which were most commonly adopted in the 1970s were MRP and its extension, MRPII [Orlicky 1975].

Materials requirements planning systems

As we have seen, the most significant drawbacks of inventory control approaches are their high cost and poor response to dynamic market changes. Manufacturing industry in the western countries in the 1970s started to undergo a fundamental change as it faced up to fierce competition from offshore countries such as Japan, Taiwan and South Korea and a staggering increase in the cost of capital, as evidenced by the rise in interest rates. During the 1960s interest rates were often as low as 6 per cent, but during the late 1970s they rose to between 15 and 20 per cent. Under these circumstances many companies looked for approaches which could increase their responsiveness and decrease their inventory levels. As a result the American Production and Inventory Control Society (APICS) became interested in Materials

24 Putting the Just-in-Time Philosophy into Practice

Requirements Planning, software was written and many companies began to use MRP systems.

The rationale of MRP systems is that they enable managers to look to the future and to increase inventory only so far as is necessary to provide for requirements which can be clearly foreseen. The mechanisms of basic MRP are as follows:

1. Forecast future demand and determine, taking the available capacity and on hand inventory into account, how much to produce to meet this demand. This process generates what is usually called a master production scheduling (MPS) that gives the production quantities of products up to a year into the future. For example, a manufacturer may decide that demand for a certain product over the following 28 weeks will be 1,000. If there are 800 in inventory and the capacity is underutilized then the MPS may indicate that 200 are to be produced in week 28. If there is insufficient rough cut capacity then less may be produced and some demand will remain unsatisfied.

   Having decided how much is to be produced, there remains the problem of ordering the material and components necessary to make the products. This is done in the second stage with the help of the Materials Requirements Planning system. This takes the MPS and breaks it down into specific raw materials and component requirements, determining the quantities of each that should be ordered, and when. As an example suppose we are concerned with making a bicycle wheel with the structure shown in Fig. 2.3 (such structures are often referred to as *Bills of Materials*).

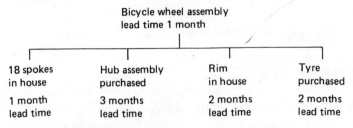

Figure 2.3 *Example of bill of materials*

The four types of component (spokes, hub assembly, rims and tyres) are assembled together to make the bicycle wheel. Of these components the spokes and rim are made in-house whilst the hub assembly and tyres are purchased from external vendors. The lead time associated with either making in-house or purchasing is given. If the MPS specifies that 200 bicycle wheels are to be produced at the end of June, then we can backtrack to determine both the ordering dates and the quantities of each component. Since it takes one month to assemble the items, all of the relevant components must be ready to assemble by the end of May. Dealing with each component in turn and taking into account the lead time:

200 × 18 (= 3,600) spokes start manufacture end April
200 hub assemblies ordered end February
200 rims start manufacture end March
200 tyres ordered end March

If there are any of the components in inventory then the amount ordered can be varied to take this into account. For example, if there are 50 hub assemblies in stock then only 150 need to be ordered. Overriding the directly computed quantities however is the minimum quantity that can be ordered. This may be given as a minimum lot size (for items made in-house) or as a minimum order quantity (for items from an external vendor). If the minimum lot size is 300 rims then 300 are manufactured and the extra 100 will remain as inventory until needed. A crucial feature of JIT is that it concentrates on reducing lot sizes and order quantities to lower the inventory levels. This aspect is discussed in later chapters.

3. Thus the output from the MRP system is a detailed list of the materials or components to be manufactured or ordered, together with the dates to start manufacture or to place orders. The activities required are to either order the components from the external vendors or to arrange manufacture. In the latter case, the task includes detailed scheduling of jobs on machines in order to ensure delivery on time. Achieving this is a tremendous problem for most manufacturing organizations.

Manufacturing resource planning: MRP II

The first MRP implementations were mainly confined to production and inventory control functions within a company. As experience was gained it became evident that the implementation would benefit if other departments were included so that the MRP implementation would cover all aspects of the company's activities, including sales, purchasing and finance. The basic MRP system was therefore expanded to encompass activities across a broader spectrum of the company's activities. This enlarged MRP system was termed manufacturing resource planning or MRP II.

Another change was evident in MRP II. The tremendous increase in computing power available had made possible a more detailed calculation of capacity usage. Prior to this, only approximate capacity planning was possible, with usage being planned across three or four very broad areas. With the increased computing power, capacity could be planned across a much larger of number of areas so that each work centre could now have a detailed capacity plan. This detailed capacity planning has been termed capacity requirements planning (CRP).

The earlier MRP systems essentially had three levels:

1. Master production scheduling
2. Materials requirements planning
3. Ordering

To incorporate the changes for MRP II three further layers were added so that the system now had six levels:

- Business planning
- Production planning
- Master production scheduling
- Materials requirements planning
- Capacity requirements planning
- Ordering

The addition of Business Planning and Production Planning broadened the range of functions covered by the MRP II process; while capacity requirements planning gave the detailed capacity analysis and provided the basis for priority planning.

Of course, in order to carry out the detailed capacity analysis accurate factory floor data are required. This increased requirement for data, in comparison with the earlier MRP systems, has caused problems in many MRP II implementations with some improvements being less than expected.

MRP problems

(The term MRP is used to refer to both MRP and MRPII systems in the following sections.)

The previous sections have briefly viewed the MRP process. The usual implementation of an MRP system is to have a manual master production scheduling stage, with a computer package to carry out the MRP breakdown into requirements for components and raw materials. This computer package requires as inputs the master production schedule, the product structure (or bill of materials) and the inventory levels. The computer package will then produce detailed requirements with the ordering dates and, in the case of MRPII, more detailed capacity analysis.

The MRP logic looks simple but a number of problems have arisen that have significantly reduced the effectiveness of many MRP operations. These problems fall into the following categories:

- Poor inventory level accuracy
- Inaccurate lead times
- Inaccurate bill of material
- Poor MPS
- Out of date data
- Poor methodology

The inventory level accuracy is crucial if accurate output is to be obtained from the MRP system. The computer software may be good in itself with very few bugs, but its performance can only be as good as the data fed into it. Most successful MRP systems probably achieve 95 per cent inventory records accuracy. If an MRP type system is implemented in an organization where inventory accuracy is lower, say only 75 or 80 per cent, then this will probably result in a large number of problems, with final products being overdue and inventory levels being kept higher than necessary.

In the majority of practical manufacturing systems lead times vary dramatically, even at the best of times. Reasons often quoted for these variations include machine breakdown, quality problems, other products taking priority, component and raw material shortages or tooling problems. My own research with a number of companies tends to bear this out; one company, for example, had a mean lead time of just a few months for one product, but this average was based on actual lead times that ranged from one month to nine months.

If we were to accept this large degree of variation, our MRP calculations (which may be based on an estimate of the mean lead times) would then become erroneous. One alternative suggested by some MRP proponents would be to have a priority, based on the due date for that particular job, attached to each item. As the due date nears, the priority would be automatically increased and then conveyed to each work centre in the form of a list of jobs (sometimes called a *dispatch list*) to be done in the near future with the urgent jobs indicated. Production could then be accelerated for the urgent jobs.

This continual updating of priorities requires substantial computing resources. The point is that although each MRP type calculation is relatively simple; the number of such calculations required can be very large. A typical batch manufacturing organization may have 15,000 types of components which combine together to form 5,000 final products. The amount of computation involved in planning, say, a week's work schedule for such a plant was, until recently, likely to need a mainframe computer running continuously over a weekend or overnight to do the MRP calculations. Although increases in both computer power and in the sophistication of the software has led to some reduction in the time required, the need for disproportionate amounts of computer time is still a problem for many MRP installations.

The bill of materials is another item of data that needs to be accurate. Inaccuracies can occur for a number of reasons: when the bill of materials file is built up in the MRP implementation, for example, or when it is altered in updating the product. The latter source of error is especially common in industries with a higher rate of product change, such as the electronics assembly business. It is generally recognized by those with experience of MRP implementations that the bill of materials should be 95 per cent accurate. But accuracy may be defined in at least three different ways:

1. The total bill of materials for each *end product* – if there is one error in the total bill of materials then it is judged inaccurate.
2. The bill of materials for each *sub-assembly* – the number of sub-assemblies with accurate bills of materials is compared with those with inaccurate bills of materials.
3. The percentage of inaccurate entries in the bills of materials.

To compare the use of these methods for judging bill of materials accuracy we can use the example shown in Fig 2.4.

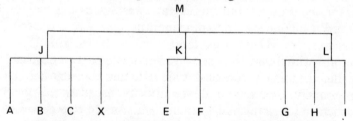

Figure 2.4 *Example of inaccuracy in bills of materials*

End product M is made up of three sub-assemblies, J, K, and L, and these sub-assemblies are made up of components A–I. If component D was inaccurately entered as X, as shown in the diagram, then the various definitions of inaccuracy (paragraphs 1–3 above) can significantly alter the percentage of error we obtain. Taking each in turn:

(a) Using definition 1 above there is an error in the bill of materials and hence the *total* bill of materials is judged inaccurate: 0 per cent accurate; 100 per cent inaccurate.
(b) With definition 2 above only sub-assembly J is inaccurate, with sub-assemblies K and L being judged accurate. The result: 66.6 per cent accurate; 33.3 per cent inaccurate.
(c) Only one entry is inaccurate, the other 12 entries are accurate. 92 per cent accurate; 8 per cent inaccurate.

Definition 2 is the one usually used, as the accuracy of sub-assemblies is a major consideration in assembly. The goal of 95 per cent accuracy of bill of materials using this definition is probably the minimum standard permissible if the performance of the MRP system is to be satisfactory. If accuracy falls below this level a significant number of incorrect parts are likely to be ordered or manufactured as a result and this is often only discovered when assembly of these components begins, resulting in a rush to complete the correct component.

The third factor leading to poor MRP performance is associated with inaccuracy in the master product schedule (MPS). The MPS acts as the driver of the MRP system. If there are significant inaccuracies in the MPS these are passed straight through the MRP system leading to errors in the MRP output. Obviously, except in the case of an industry that relies on firm orders being received before ordering raw materials and commencing manufacture, the MPS, which may determine the production rate up to one year ahead, will usually contain a degree of inaccuracy. But the point is that inaccuracy increases with the length of the forecast; in other words, the predicted demand for next week, for example, is likely to be more accurate than that for the month after next. Thus, if the raw materials and manufacturing lead times are reduced then the period over which it is necessary to produce forecasts will also be cut.

Accepting that the MPS is likely to be inaccurate to a greater or lesser extent, the problem becomes one of reducing the damage caused by such inaccuracy. One way to reduce any error is, as already mentioned, to reduce the lead times. Another is to set a freeze period within which nothing on the MPS can be altered. For example, it may be laid down as a rule that the MPS for the next eight weeks must always be left unchanged. Many JIT implementations use this technique when ordering from a supplier. Harley Davidson, for example, in their JIT approach (which was one of the earliest JIT implementations in the USA) frequently used a six-week freeze period when ordering from suppliers. This gave the suppliers the advantage of knowing that they could go ahead to complete the orders with the guarantee that they would not be changed.

Most experts in MRP implementation would suggest that, again, 95 per cent accuracy in the MPS is essential to the MRP operation. This accuracy is based on 95 per cent of the MPS being correct – in *number of items*, not in money. Reasons for an inaccurate MPS include inaccurate forecasts of demand and inaccurate capacity data. As already indicated, both these elements tend to be inaccurate in practice since, first, it is difficult to accurately forecast the demand for most items any distance into the future and, second, capacity usage is dynamic, changing rapidly from one time period to another. Under these circumstances it is extremely difficult to judge levels of capacity availability in the future.

The other difficulty for MRP systems arises from out of date data and the time it takes for data to be entered into the system and become accessible. This time delay can be due to two reasons:

1. The low frequency of computer runs, which often take place only once a week. This means that data are out of date for much of the week.
2. The time delay between some transactions taking place and the record of these transactions being entered into the computer.

In the past, MRP computer runs have usually been organized during off-peak times, often at weekends, since the computing resources required were substantial. With recent improvements in both the computer hardware and the MRP software, the computing times associated with the MRP runs have been reduced so that many MRP users now achieve overnight runs. With a weekend run the data were out of date by Wednesday; with an overnight run the data are only a few hours old. This greatly increases users' confidence in the output of the MRP system. In many companies which operate MRP that I have worked with, the practice of only having an MRP run once a week has been a continual problem. By the latter part of the week the MRP output can be extremely inaccurate which in turn can result in diminished confidence in the MRP system. This sets off a vicious circle of decline along the following lines:

- Something causes loss of confidence in the MRP operation. This may be inventory or bill of materials inaccuracy, out of date data or MPS inaccuracy.
- This loss of confidence gives users less incentive to take the necessary steps to increase the accuracy of data input

- the further decline in data accuracy causes a further erosion in the quality of the MRP output and a further loss of confidence.

The second and third steps are repeated cyclically, resulting in decreasing MRP effectiveness. I was once a consultant to a company where this had happened. The MRP system had been poorly implemented and the inventory, bill of material and MPS were all inaccurate. The resulting problems had caused the MRP system to become more or less unused. The MRP runs were still done but no one had any faith in the output, relying instead on an informal manual system. Work in progress and inventory levels had escalated and morale was very low. When I was called in I recommended a reimplementation of the MRP system together with changes to ensure inventory, bill of material and MPS accuracy and some changes to take advantage of JIT. The MRP system was reimplemented, and the other changes were introduced. The result was a successful MRP operation which could then form a firm basis for the move towards JIT.

The last problem area associated with MRP implementation concerns poor methodology, and this raises the question: is the MRP methodology flawed, especially when used for shop floor control? In order to answer this, let us go back and look at the major features of the MRP system. First, we forecast demand, then take the capacity into account to determine the MPS. The MPS is put into the MRP system together with the inventory and bill of material records and the component/raw material requirements are produced. If we examine, as we have briefly in this chapter, each of these features then we quickly see that each stage involves some error being placed into the calculation. What should end up as a highly accurate requirement for raw materials and components together with a highly accurate schedule of deadlines often contains considerable errors.

The shortcomings of the MRP methodology have been most evident when it has been used as a centralized facility to schedule the day-to-day activities in a factory. This requires the provision of accurate, detailed data relating to every aspect of the factory's operations and when MRP systems are expected to do this detailed scheduling the results have been disappointing. MRP systems should be used where they are most suitable; that is, for coordinating different parts of the factory. In the future we are likely to see MRP systems being used to keep global control of factories while detailed control will be achieved by JIT approaches. The ways in which MRP and JIT can be combined, each being used for the tasks to which it is best suited, are explored in later chapters.

3
JUST-IN-TIME OVERVIEW

Introduction

As will be stressed throughout this book, the average manufacturing operation is extremely complex to manage. The work of several hundred people and dozens of machines has to be coordinated to produce hundreds of product types, and it is not surprising that managers have frequently failed to control such a complex operation effectively.

The traditional approaches to manufacturing management, such as MRP, rely on a fairly strictly defined methodology which supplies detailed figures to managers on what should be produced [see Fox (1982), Garwood (1984) and Swoyer (1983)]. As was indicated in the previous chapter, the results from these traditional approaches have been disappointing, in spite of enormous sums of money having been spent on putting them into practice. The average MRP II user has probably spent over $1 million on its implementation [see Wallace (1985)]; in addition, considerable amounts of valuable management time will also have been devoted to the implementation.

This chapter gives an overview of JIT. The basic points behind the philosophy are outlined since these provide the framework around which the five steps necessary for successful implementation can be built and a good understanding of the JIT philosophy is essential to fully appreciate what is involved in each of the five steps.

The first two sections draw a distinction between the traditional approaches and JIT and prepare the ground for a discussion of the four principles behind the JIT philosophy. Each of these four principles is described and examples are given of how they impinge on a typical JIT implementation. Any person who has a firm grasp of these principles will have assimilated JIT philosophy. Such an appreciation of JIT philosophy leads us into the final section of this chapter which describes the costs and benefits of implementing JIT.

Overall the chapter provides the backdrop against which the five steps for implementation can be placed.

JIT is not a software package

Almost all other approaches to manufacturing management revolve around some sort of software package. The software is purchased along with the necessary computer hardware and peripherals and, provided that the correct data are fed into it (this can be a big problem, as we saw in the section on materials requirements planning in Chapter 2), then an answer is produced which provides the basis for managerial action.

Examples of software packages for manufacturing management include:

1. Optimized production technology (OPT) [Fox (1982)]. OPT is an approach which takes the view that production is oriented around bottleneck processes; these bottlenecks limit the flow through the whole manufacturing system. Although several associated factors are stressed, including the requirement for accurate data input, OPT revolves around a computer package that produces what are hoped to be accurate schedules which can then be diligently followed by the managers and supervisors.
2. Materials requirements planning (MRP and MRPII). The MRP and MRPII systems operate around a computer software package that produces detailed requirements (with dates) for sub-assemblies and raw materials.
3. Reorder level/reorder point/reorder quantities. Although these approaches are simple enough to be operated manually for a small manufacturing operation, they would benefit from being computerized.

These three approaches essentially follow the same pattern:

- data is collected
- it is then fed into a computer package
- the output is used as the basis for managerial action

JIT does not operate like this. Although data analysis does form part of the operation of a typical JIT installation this aspect is not viewed as a central issue. Instead of an enormous amount of data analysis which is typical for OPT, MRP or ROP/ROL/ROQ approaches, managerial action is concerned with more fundamental steps that ensure that work goes smoothly through the manufacturing system. Where OPT, MRP, etc. are data *driven* – that is the data analysis is the major input for managerial decision making – with JIT the management is mainly concerned with creating the right environment for effective operations.

The first environment area is *strategic*, this covers the major, fundamental issues that govern the operation of the company. Examples are the choice of products to be manufactured, the control mechanisms for the factory and the cost of production, including set-up time, scrap and other quality costs.

The second area is *tactical*, this is concerned with actions and decisions that have a relatively small impact on the operation of the company. Examples are deciding which job has the highest priority on a particular process, or determining the amount to be ordered from an external supplier.

OPT, MRP, ROP/ROL/ROQ type approaches concentrate on the tactical level, and can give an indication to a manager of the actions to be taken at a detailed level. But they do not have any effect on the major decision-making areas which are strategic in nature. The JIT approaches move managers away from the detailed tactical decision-making towards the more strategic areas such as reducing the product range via more standardization, reducing set-up time, decreasing scrap rates, etc. This means that the attention of managers and supervisors is directed towards those areas that can most benefit from their efforts and away from the minute-by-minute, instantaneous response type of management.

To take an example, there is a traditional textile plant in North Carolina which, until recently, had the usual problems associated with the vast majority of manufacturing companies; that is, long manufacturing lead times, high work in progress, high reject rates, etc. The production manager worked 12-hour days, spending nearly all his time 'firefighting'. When a problem arose, as it did several times each hour, he would hurriedly investigate, make a few phone calls and take some corrective action. The problems were typically tactical in nature. Six months after implementing JIT, the plant had much reduced manufacturing lead times, lower levels of work in progress and low reject rates. The production manager spent his (reduced) working week on the *strategic* areas of concern; improving process reliability, reducing scrap rates, etc. His role had been changed from that of a firefighter to more of a foundation builder – he became concerned with doing the ground work that would allow the company to achieve greater operating effectiveness. The result is that the company has become more profitable and looks forward to a future in which profitability will continue to increase. The manager is also far happier in his job.

In short, an effective JIT implementation will mean that instead of facing a day filled with a succession of crises, each requiring a decision to be made, production managers, supervisors and everyone else in the company will have the time to stand back and resolve the fundamental issues which can move the company to greater effectiveness and greater profitability.

JIT is not a methodology

The other major manufacturing management approaches are based on a fairly rigid methodology. Data are fed into such systems, defined operations are performed on the data, and the manager is supplied with some answer which will form the basis for his actions. By contrast, JIT is a fairly loosely defined technique. To implement JIT we do not use formulae with complex derivations. Gone are the ideas behind, for example, the economic order quantity which determines the 'optimum' amount to be ordered. Instead we replace the rather rigid methodological approaches by much more flexible ones that are not intended to obtain 'optimum' solutions to small-scale, tactical problems but instead to resolve some of the fundamental problems.

If, for example, we examine the use of the economic order quantity then we will find that something like 1,000 variations have been derived. These variations address both the quantity to be ordered from an external supplier and the quantity to be made in-house (sometimes called the batch size). They cover a wide range of permutations including random demand, manufacturing lead times, etc. But as the complexity of the problem grows so does the complexity of the formulae.

When JIT is implemented, instead of just using a formula we examine all the assumptions of the operation. If we look at the question of batch sizes, we can ask ourselves, What is to prevent us reducing the inventory cost by reducing the batch size? The answer is, The set-up cost. Reduce or eliminate set-up and we reduce both the batch size and the total cost of the operation. But in order to achieve this we must tackle fundamental problems which no formula can solve.

JIT therefore marks a substantial shift away from the other approaches to manufacturing management. Instead of purchasing a software package and/or using a formula to obtain a definite solution, the manager who adopts the JIT approach must examine and gradually evaluate the fundamental problems of the organization. Whereas a manager could previously hide behind a software package and a well-defined methology, the JIT manager has to deal ruthlessly with the inefficiencies that are holding the organization back.

JIT is, therefore, best defined as a *philosophy* that, when implemented correctly, will permeate every section of the company and change the way in which everyone operates.

36 Putting the Just-in-Time Philosophy into Practice

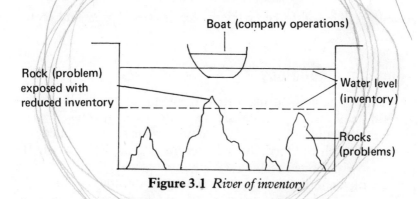

Figure 3.1 *River of inventory*

Attack fundamental problems

The Japanese culture is one which likes to represent concepts pictorially. To describe the first objective of the JIT philosophy, attacking the fundamental problems, the Japanese use the analogy of the *river of inventory* (see Fig 3.1). The level of the river represents inventory and the company operations are visualized as a boat which navigates up and down the river. When a business attempts to lower the level of the river (in other words to cut its inventory levels) it uncovers rocks, that is, problems. When such problems arose in industries in western countries the response, until fairly recently, was to put increased inventory in place to cover up the problem. A typical example of this kind of problem would be a plant which contained an unreliable machine feeding parts to another, more reliable, machine – and the typical response of western management would be to keep a large buffer stock between the machines to ensure that the second machine did not run short of work. By contrast, the JIT philosophy indicates that when problems are uncovered they must be confronted and solved (the rocks must be removed from the river bed). The inventory level can then be gradually reduced until another problem is uncovered; this problem will then be attacked, and so on. In the case of the unreliable machine, the JIT philosophy would indicate that the problem should be resolved either by a preventive maintenance programme that would improve the machine's reliability, or if all else failed, by the purchase of a more reliable machine. This difference between traditional western and JIT approaches is illustrated in Fig. 3.2.

Some of the other problems (rocks) and JIT solutions are shown in Table 3.1. Where there is a machine or process that is a bottleneck, one of the traditional western aproaches has been to aim for better and more complex scheduling (using, for example, MRP II) to ensure that it

Just-in-Time Overview 37

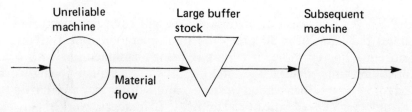

Traditional Western Approach

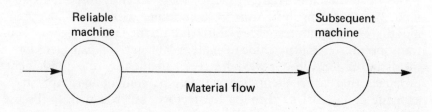

JIT Approach

Figure 3.2 *Approaches to unreliable machines*

never runs short of work, so reducing the effect of the bottleneck. The results of such policies have frequently been disappointing – stockturn objectives, which are a good measure of efficiency, have been lower in western countries than in Japan, and, what is more, stockturn objectives have been increasing faster in Japan than in western countries. The JIT approach to the presence of a bottleneck machine or process would involve reducing set-up time to produce greater capacity, finding alternative machines or processes, purchasing extra capacity or even subcontracting excess work. A JIT manager recognizes that neither increased buffer stock nor more complex scheduling will solve the fundamental problem; all they do is to temporarily cover over the rocks.

Table 3.1

PROBLEM (ROCKS)	JIT SOLUTION
Unreliable machine	Improve reliability
Bottleneck areas	Increase capacity
Large lot sizes	Reduce set up time
Long manufacturing lead times	Reduce queuing, etc. by using a pull system
Poor quality	Improve processes and/or supplier

Many typical western manufacturers will have manufacturing lead times that are 20 or 30 times the total operation time – if they are efficient! For example, if a part is being manufactured with a total operation time (that is, the time the product is actually being worked on) of 10 hours, then a good western manufacturer may well have a total manufacturing lead time of 250 hours. If 80 hours are worked per week, then this is just over 3 weeks. It should be stressed that this is a good manufacturer; many have lead times far in excess of this. One company I know had a total operation time for its product of 16 minutes and a total manufacturing lead time of 6 months, with items spending 6 months, less 16 minutes, either cluttering up the factory space or being transported from one machine to another. When a JIT manager looks at the issue of long lead times, he seeks to identify the fundamental problems that cause them. He will not be content to simply try to expedite certain orders through the factory, but will seek to discover why the manufacturing lead times are so long. In my experience long manufacturing lead times are the result of a number of factors, including bottleneck machines or processes, poor reliability of machines, poor quality control (this requires below-par items to be reworked – a very expensive business), and poor shop floor control. By attacking all of these problems, lead times can be gradually reduced.

(2) Eliminate waste

The second objective of the JIT philosophy is expressed in a phrase frequently used in the more efficient Japanese factories – eliminate waste. Waste, in this context, means everything that does not add value to a product. Examples of operations that add value are the processing operations such as metal cutting, soldering, electronic component insertion, etc. Examples of operations that do *not* add value are inspection, transport, storage, setup.

Take the case of inspection and quality control as an example. The traditional western approach is to have inspectors strategically placed to examine parts and, if necessary, fail them. This has a number of disadvantages, including the time it takes for parts to go through the inspection process and the fact that the inspectors often discover faults only after a whole batch has been manufactured, which necessitates the whole batch either being scrapped or reworked, either of which will be very expensive.

The JIT approach is to eliminate the need for a separate inspection stage by emphasising two imperatives:

1. Make it right the first time. Since achieving high quality items usually involves no greater expense than producing low quality items, we might as well produce high quality. All that is needed is a concentrated effort to iron out any defect-producing trends.
2. Make it part of the operator's responsibility to monitor the process and take corrective action where necessary. This can be done by giving them guidelines that they should aim to achieve.

If we compare the traditional western approach to inspection and quality control with the JIT approach (as shown in Fig 3.3), we see that the traditional western approach has been to set upper and lower bounds (for example, tolerances) and if the measurements fall outside either the upper or lower bound, then the item is scrapped or reworked.

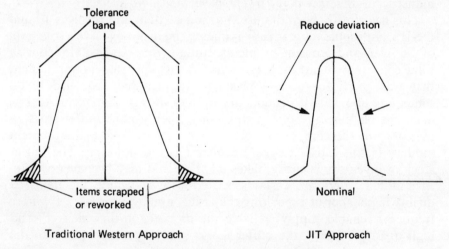

Figure 3.3 *Approaches to tolerances*

By contrast, the JIT approach is to reduce the deviation from the ideal nominal, any deviation from the nominal not being tolerated. Furthermore, JIT places the responsibility for detecting and correcting deviations for the nominal on the operators of the processes. They are expected to make it right the first time and to prevent *any* items from deviating too far from nominal. These are the essential features of statistical quality control.

Storing inventory is another example of a wasteful activity. The real cost of inventory is two-fold. The first cost is, of course, the direct cost in terms of capital and warehouse costs and the risk of stock becoming obsolete. Most industries in western countries have been assuming this cost of inventory to be somewhere between 20 and 30 per cent per year, so that a $150 item in stock for a year would cost between

$150 × 20% and $150 × 30% or between $30 and $40.

There are some variations on this. I worked with a company recently where the cost of inventory was taken to be 4.7 per cent per year! This was a figure computed in the late 1960s when interest rates were extremely low and it had never been changed. This particular company had large amounts of inventory, in raw materials, work in progress and finished goods and seemed very happy with what they thought was the low cost of inventory. A quick examination of the 4.7 per cent was enough to persuade them to change to a more realistic figure which highlighted the need for action to reduce the inventory levels.

The second cost, which has been overlooked in traditional western industries, is that *inventory keeps problems hidden from view*.

This has already been discussed to some extent and revolves around the JIT philosophy of gradually reducing the inventory levels (the level of the river) and exposing problems. Initially this idea was viewed with some concern by many managers who felt that if a problem was hidden, that was good; for once a problem was exposed they had to do something about it. If vendors are not delivering components or raw materials on time and of a high quality, then a large buffer stock of inventory of the components or raw materials will keep the problem hidden. Is this satisfactory? The answer is a definite *no*. The buffer stock of inventory is costly, takes up space and may become obsolete. In addition, there are costs associated with returning the items (if quality is poor) or of expediting the order (if delivery is late). By more frequent, reliable supply (reliable means both on-time delivery and high quality supply), the buffer stocks can be reduced and also the costs.

Eliminate waste, therefore, is a phrase used to direct the attack. Eliminating all activities that do not add value to a product reduces costs, improves quality, reduces lead times and increases customer service levels. Indirectly, of course, this also can increase sales. The company in case study 2 (Chapter 9), for example, doubled their sales using JIT approaches.

Eliminating waste involves much more than a single once-and-for-all effort. It requires a continuing struggle to gradually increase the efficiency of the organization and demands the cooperation of a very large proportion of the workforce of the company. If the policy is to be effective it cannot be left to an 'eliminate waste committee' but needs to permeate every corner of the company operations. It can become the rally cry or the slogan of a JIT implementation. But, of course, it will be effective only if the concepts are fully understood by the employees and only if action is taken to implement the eliminate waste strategy.

Companies who have achieved the most out of using the eliminate waste philosophy have generally:

- not viewed it as the sole criterion
- have mobilized the employees to fully implement the philosophy

Suggestion schemes are one way to involve the employees. In the past, such schemes have been notorious in the West for the poor employee response which has been obtained and for the low proportion of suggestions which have actually been implemented. I worked with one company which had several thousand employees and each year a $500 prize was offered for the best suggestion to improve productivity. The suggestions that did not win the prize were never implemented, and even the one suggestion that won the prize was frequently not used. What was the result? Every year there were fewer and fewer entries. The main reason for this was the low implementation rate – employees were in effect being told that their ideas were not worth considering.

But just one example will serve to show the value of shop floor expertise – when it is recognized and properly exploited. In another company I worked for we were producing some pressed steel items for use in power switches. A large (and very profitable) order came through, but it had a very short lead time associated with it. Unfortunately we had many problems getting the steel press to stamp out one particular component. Every day we would try something new and each time the pressed steel would come out of the press and spring out of shape. We would then go to the press operator, George Boothroyd, and tell him to alter the thickness of the steel, or to alter the steel type, or to use higher tolerance dies, and still the items would spring out of tolerance. One day, after all these had been tried to no avail, George said to me, 'The answer is simple, really.' I was absolutely floored. 'What do you mean, George?', I said, and then George proceeded to tell me how to alter the die so that the item sprung *into* tolerance. We tried this and it worked perfectly. Afterwards, I said to George, 'Why didn't you tell us earlier about this?', and George said, 'Because you never asked me!'

What is the moral of this story? Basically, that people who spend their time working on machines or processes are likely to have a very good understanding of them. They can make an important contribution to improving the process and achieving the goal of eliminating waste. Suggestion schemes can be valuable if the suggestions are used. The acceptance rate in the majority of factories in the better Japanese companies is very high (the average is probably somewhere about 80–85 per cent), and this, in turn, raises the morale of the employees and results in an increased number of suggestions being received.

If waste is to be effectively eliminated, the programme must involve the full participation of the majority of employees. This means that the traditional approach, based on telling each employee exactly what he or she has to do must be altered, and the JIT philosophy stresses the need to respect workers and to include their input when formulating plans and operating the plant. Only by doing this can we utilize the expertise and experience of all employees to the full.

This more participative management style can demand considerable adjustment especially on the part of supervisors and foremen. Personnel at this level can often feel that their power base is being eroded if they are not kept fully informed of the purpose of the changes involved in a JIT implementation. However, I have found that if the personnel of the company, especially at the supervisor/foreman level, are fully educated about JIT, then it is likely that the JIT approach will command their wholehearted support.

Strive for simplicity

The third objective of the JIT philosophy is to strive for simple solutions.

The approaches to manufacturing management that were in vogue during the 1970s and early 1980s were based upon the premise that complexity is unavoidable. On the face of it this is a fair assessment: a typical batch manufacturer may well have several hundred batches at any one time passing through dozens of processes. Each batch is likely to involve a number of separate operations and will probably travel over much of the factory floor before it is completed. The complexities of managing such a system are phenomenal; the interactions between jobs as well as the requirements for other resources usually overwhelms most managers.

JIT stresses the desirability of simplicity on the grounds that simple approaches are most likely to lead to more efficient management.

The primary thrust of the drive for simplicity covers two areas:

1. Material flow
2. Control

The simple approaches to material flow aim to eliminate complex route paths by moving towards more direct, if possible unidirectional, flowlines. As mentioned earlier, the majority of batch manufacturing plants are organized in what may be termed a process layout; a typical example being that shown in Fig. 3.4.

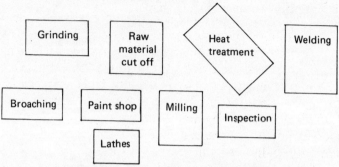

Figure 3.4 *A typical process layout*

Most items going through such a factory would follow a tortuous route going from, for example, raw material cut off to lathes then to broaching, welding, grinding, heat treatment, grinding and paint shop. Each process usually involves a considerable amount of waiting added to which is the time taken (amid the general confusion of shop floor activity) to transport items from one process to another. The consequences of this are well known: high work-in-progress levels and long manufacturing lead times. The problems of trying to plan and control such a factory are phenomenal, with typical symptoms being late items rushed through the factory whilst others, no longer being immediately required because of a cancelled order or a change in forecast, are stopped and remain stagnant on the factory floor. These symptoms have very little to do with the effectiveness of the management. No matter how good a manager is, he is going to have problems in controlling such a system. Now we can try to cope with the problem by, for example, putting in a shop floor computer control system but if the factory still remains tremendously complex then the benefits obtained will probably be marginal.

The JIT philosophy of simplicity examines the complex factory and starts from the view that there is little to be achieved by placing complex control on top of a complex factory. Instead JIT stresses the need to simplify the complex factory and to superimpose a simple system of controls.

How is the simple material flow in the factory achieved? There are a number of ways of doing this, the majority of which can be implemented simultaneously and these are examined in more detail in Chapters 6 and 7. Briefly the main method is to group the products into families, using the ideas behind group technology, and rearrange the processes so that each product family is manufactured on a flow line. The ideal case is shown in Fig. 3.5, although, as described in Chapter 6, the flow line is often arranged in a U shape.

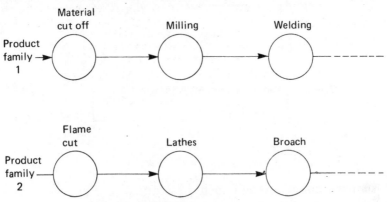

Figure 3.5 *Product layout using flow lines*

Items in each product family can now move from process to process more easily as the processes are placed adjacent to one another, hopefully thereby reducing work in progress levels and manufacturing lead times.

With these small flow lines in place, other advantages follow. For example, the management becomes much easier than in the case of a process layout since each flow line is, to a large extent, separate. A sub-manager can be responsible for each flow line. In addition quality will tend to be improved; since there is less likelihood of panic with fewer rushed orders, more time can be spent addressing the quality problems.

The JIT philosophy of simplicity, besides applying to the flow of items, also applies to the control of these flow lines. Instead of complex control (along the lines of MRP and OPT), JIT places the emphasis on simple control. An example is the pull/Kanban system. This is described in more detail in Chapter 7, and marks a significant departure from conventional control approaches. These are designed to *push* work into the factory; in contrast, the pull/Kanban system *pulls* work through.

MRP (and OPT) are push systems in the sense that they plan what is to be produced which is then pushed through the factory. Bottlenecks and other problems are supposedly detected beforehand and complex monitoring systems are installed to feedback factory floor changes so that corrective action can be taken. By contrast the JIT approach using the pull/Kanban system eliminates the complex set of data flows since it is essentially, in its original form, a manual system. When work is taken from the last operation, a signal is sent to the preceding operation to tell it to produce more items; when that operation runs short of work, it in turn sends a signal to its predecessor, and so on. This process continues backwards through the manufacturing flow line as shown in Fig 3.6.

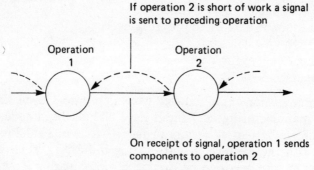

Figure 3.6 *The operation of a pull system*

In this manner work is pulled through the factory. If no work is taken from the end operation no signals are sent to the preceding operations and hence no work is done. This is a major departure from previous approaches to material control. With reduced demand, personnel and machines do not produce any items. JIT proponents suggest that they do other tasks such as cleaning the machine, adjustments, checking for any required maintenance, etc. With more traditional approaches, most managers are less keen to have personnel and machines remaining idle. Work would be scheduled even if it was not required in the near future. All too often it was never needed because the product had become obsolete and the completed items had to be scrapped. In effect, the traditional approach attached the highest priorities to keeping machines and personnel producing, even at the cost of manufacturing items that only contributed to an already swollen inventory and a high scrap rate.

The JIT approach, based on the use of pull type systems, ensures that production does not exceed immediate needs, thereby reducing work in progress and inventory levels as well as cutting manufacturing lead times. The otherwise idle time can be productively spent removing the source of future problems by a preventive maintenance programme. To achieve the correct environment for this to occur requires a comprehensive education, training and communication programme as described in Chapter 5. The evidence from western manufacturers who have implemented such a programme shows encouraging results in reducing manufacturing lead times and machine down time. In addition, morale has been significantly raised.

There are a number of major benefits to be gained from the use of the JIT Kanban/pull type systems, including the following:

- reduced work in progress levels
- reduced inventory levels
- reduced manufacturing lead times
- gradual reduction of work in progress levels
- identifying bottleneck areas
- identifying quality problems
- simple management

The first three advantages of reduced work in progress levels, inventory levels and manufacturing lead times have already been covered. The original Kanban system at Toyota in Japan achieves a stockturn of 80 compared with the western manufacturing industry average of about 3–4. Although care needs to be exercised in comparing Toyota's line with that of other industries, the figures do indicate the staggering advantages that Kanban/pull type systems can bring.

One of its major advantages is that they simplify management of the manufacturing system. Previously a manager had to try to control and coordinate a whole manufacturing system with a process type layout as shown in Fig 3.4. Admittedly, the manager had at his disposal the sophisticated and highly complex information produced by a central planning computer but this brought its own problems. By contrast, the pull type systems run themselves, with a much reduced need for complex computer control. The constraints of the system rather than the output of a computer determine the flow of work. If, for example, a bottleneck area develops it will slow activities upstream too so as to prevent a pile-up of work in front of the bottleneck.

The improvements associated with pull type system occur in a gradual manner. They seem to work best if the system is first set up with fairly large queues in front of each process and the level of the river (the work in progress) slowly reduced to lessen manufacturing lead time. The improvement may be slow, but it will also be continuous; Toyota, for example, had been gradually improving their Corolla line for over 20 years before achieving the stockturn figure of 80, and they are still improving the line to obtain stockturn approaching 100. It is often thought that pull/Kanban systems can only be used where there is little product variety and little demand variation. However, as described in later chapters, many companies are using adapted pull/Kanban systems where these conditions do not apply.

The fact that pull/Kanban type systems identify bottlenecks and

Just-in-Time Overview 47

other problems was initially looked upon as a disadvantage in the west. Why should we want to identify problems? Why not leave well alone? Well, as indicated earlier, JIT aims to solve the fundamental problems, and this can only be done if the problems are identified – the fourth plank of the JIT philosophy.

Devise systems to identify problems

We saw how pull/Kanban systems bring out problems. Another example is the use of statistical quality control which helps to identify the source of the problem. Under JIT, any system which brings out problems is considered beneficial and any which masks problems, detrimental. Pull/Kanban systems identify problems and are thus beneficial. Previous approaches tended to obscure fundamental problems and thus delay or prevent their solution. For example, if there was a long set-up time then the use of the EOQ led to large batch sizes; but there was no mechanism within EOQ which ensured that the user got the message that the setup times were too long. Many other problems also occur in the majority of manufacturing systems;

- unreliable suppliers
- poor quality
- bottleneck processes, etc.

The systems set up within a JIT implementation should be designed so that they immediately trigger some warning whenever a problem arises. Once again we can look to the Toyota Corolla production line for an example. On this line each worker is allocated parts on a one-to-one basis; that is, if each car requires one centre stoplight then the worker is only allocated enough stoplights to cover the number of cars produced. If there is a fault with one of the stoplights then the worker is forced to draw attention to it. He does this by pulling on a rope that is dangling above his workstation and when the rope is pulled the entire production line stops – something, it can readily be imagined, that guarantees that the problem is brought to the manager's attention! This dramatic step is followed by a hurried conference of all concerned where every effort is made to solve the problem. This whole process can be described diagrammatically as shown in Fig. 3.7. The line is operating at a certain efficiency, a problem occurs, the line stops and the problem is identified, some remedial action is taken and later the line is restarted. Since a problem has been confronted and wholly or partially solved the restarted line is less likely to suffer from that particular problem again and the efficiency of the line is thus increased.

The vertical scale in Fig 3.7 has been exaggerated to show the effect but it can easily be seen that the approach works by the gradual accumulation of a series of small increases in efficiency. When enough of these small increases are combined, the result is a major increase in efficiency. This has been the experience of the Toyota Corolla line.

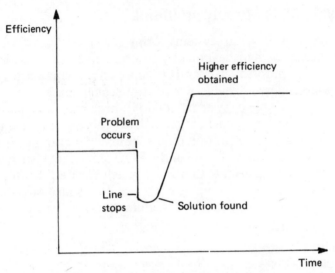

Figure 3.7 *Incremental efficiency increase*

We can use some of these ideas in any JIT system we design. The aim is not only to lower work in progress levels and manufacturing lead times but also to identify problems as soon as possible after they occur and to force managers to take remedial action.

For example, if we have a bottleneck process then ingenious scheduling may alleviate the symptoms but it will never remove the problem. In fact, more complex scheduling simply skirts around the problem at the cost of, for example, keeping extra work in progress, rescheduling work into other less efficient processes or altering the product mix. Even worse, it also serves to obscure the problem; for a manager may still be able to schedule a factory with several bottlenecks without being forced to recognise that his operation has built-in problems which should be identified and solved. In order to identify a problem properly, a manager should be prepared to pay the price in terms of some short-term setbacks. in the hypothetical example given earlier the entire Toyota Corolla line was stopped because of a fault in a single centre stoplight; this is something many managers who take a short-term view would consider unthinkable. In the long term, however, it is often only by ensuring that even a small problem has a major impact that attention will be concentrated upon it.

If we are really serious about implementing JIT we then have to do two things:

1. to set up mechanisms so that problems can be identified
2. to be prepared for a short-term drop in efficiency to gain a long-term advantage

Having spent many years in manufacturing control and visited and worked with many manufacturing companies I am quickly able to sum up the quality of the manufacturing control. A quick look at the factory floor shows the presence (or absence) of bottlenecks, the overall level of work in progress gives a good idea of the quality of the management and the morale on the factory floor can be readily discerned. The difference between a typical traditional company and a JIT implementation is marked. The low work in progress levels in a good JIT implementation give an uncluttered almost deserted feel to the factory floor. The higher morale of a typical good JIT implementation is also noticeable, with a more dedicated atmosphere prevailing.

This fourth and final aspect of JIT philosophy – the need to set up systems to identify problems – can initially be viewed as a potential disadvantage by many managers. However, the evidence does show that if systems are set up to identify problems and if they are then solved (this being the first aspect of JIT philosophy) the operation of the company can be significantly improved.

Cost/benefit of implementing JIT

The conventional approaches to manufacturing control such as MRP or OPT require large injections of capital. For example, implementing MRPII is estimated to cost each company an average of over $1 million [see Wallace (1985)]. Much of this cost is in computer hardware and software. Typically an implementation of MRP or OPT involves an eighteen-month implementation sequence to solve the data flows; the system is then tried out in parallel with the existing system, initial troubles are remedied and finally the company is switched over to the new approach.

By contrast, JIT involves little capital expenditure. What is required is a reorientation of people towards their tasks. That is one reason why this chapter has stressed the importance of understanding the JIT philosophy.

With JIT implementation any costs involved are mainly costs of education. The personnel within a company have to be aware of the philosophy behind JIT and how that philosophy impinges on their own particular function.

But while the cost of a JIT implementation is lower than that of a typical MRPII implementation, the reduction of inventory with JIT is potentially much greater, many implementations resulting in inventory reduction of 60–85 per cent. We must also remember that JIT should not be considered in the short term; that is, we should not have a JIT campaign for a six-month period and then stop. Instead, JIT is an ongoing campaign – we are striving for continual improvement. We should also bear in mind that JIT does not only reduce inventories but also increases quality, customer service and the overall morale in the company.

All the evidence demonstrates that JIT can give a good return on investment, the one proviso being that the implementation is well-planned and carefully executed. The major steps towards successful implementation are discussed in later chapters, and the chances of a good implementation can be further increased by expert advice, education and intelligent management.

Conclusion and summary

This chapter has provided an overview of JIT systems. The first part of the chapter showed how JIT marks a significant departure from more traditional approaches. The section entitled 'JIT is not a software package' points out that JIT does *not* involve purchasing a software package that provides detailed operational answers.

The following section, 'JIT is not a methodology', stresses that JIT is not a computerized black box into which we can feed data and obtain solutions. Instead JIT is a philosophy with four main pillars:

1. *Attack fundamental problems.* Instead of continually firefighting, a JIT manager will spend more time solving some fundamental problems. By solving enough fundamental problems the root causes of firefighting are removed and management becomes much more effective.
2. *Eliminate waste.* This covers a wide variety of activities but all are associated with the eradication of activities that do not add value to a product.
3. *Strive for simplicity.* The average manufacturing company is an extremely complex operation. The traditional approaches to manufacturing management have usually tried to superimpose

some sort of complex control upon the complex operation. These approaches are doomed to fail. By contrast, JIT sets out to simplify the operation of the company. If this is accomplished, simple control schemes can be used to control the simplified system.
4. *Devise systems to identify problems*. Before fundamental problems can be removed, they must be identified. JIT stresses that systems should be used to draw attention to the problems. Examples of such systems are the pull/Kanban system used for JIT shop floor control and statistical quality control.

The principles behind JIT philosophy are illustrated by the analogy of the river of inventory, in which the operation of the company is equated to a boat navigating a river. The level of the river is equivalent to the level of inventory. As we *gradually* reduce the inventory level, rocks are exposed. These rocks can then be removed so that gradually the level of the river can be lowered.

The final part of the chapter describes how JIT can be a low cost/high return policy. A successful JIT implementation can achieve good benefits for a company with relatively low costs. However, any successful implementation will only occur when the four principles of the JIT philosophy described in this chapter are fully understood. Only then can the five steps described in the remainder of this book can be started.

4
STEP 1: GETTING THE BALL ROLLING

Introduction

The four principles discussed in the previous chapter provide the framework around which we can formulate five major steps required for a *successful* JIT implementation – by *successful* we mean an implementation that is professional, educated and determined and which yields the greatest possible benefits.

The first step is to 'get the ball rolling', and the way in which this is done is crucial, as the decisions taken at this stage will in large part determine whether the implementation will be accomplished in a professional fashion or whether it will be a short-sighted, half-hearted affair.

Even a poor implementation will produce some benefits; but to do the job well costs no more than to do it badly and there is everything to be gained by tackling the job properly from the start.

'Getting the ball rolling' involves a number of stages which taken together, should set the company on the road to a successful implementation. These stages include:

- Basic understanding
- Cost benefit analysis
- Commitment
- Go/no go decision
- JIT project team selection
- Pilot plant identification

Basic understanding requires that a few key personnel have a thorough grasp of the JIT philosophy. These pathfinders, having seen the benefits to other companies who have already implemented JIT, will be the driving force behind the acceptance of the JIT philosophy in their own company. They will be involved in the *preliminary education* of the senior decision makers in the company, organizing intensive seminars and briefings that will provide enough information to assess the possible benefits and pitfalls.

Cost/benefit analysis then follows, using other implementations as a guide if possible, and this leads to a *Go/No Go decision*. If the company commits itself to JIT, a project team is selected and the pilot plant identified.

It is essential that each of these stages is conducted in a thorough and professional manner. This chapter describes them in detail and gives some guidelines to help ensure that they lead to a successful JIT implementation.

Basic understanding

Most companies have a small number of extremely valuable employees; those with the foresight and initiative to investigate new approaches to manufacturing that could alter the company's future. These vital staff, who we shall call the *pathfinders*, look to the future and promote innovation within the company. They can have a variety of job functions, but each 'pathfinder' is identified by the fact that they consider the long-term welfare of the company beyond their day-to-day responsibilities. Many JIT implementations begin with a few 'pathfinders' having heard about JIT, from talking to employees of other companies, a professional society or at a seminar or conference. They investigate further under their own initiative. Without such 'pathfinders' the company is less likely to be aware of new developments and therefore less likely to retain its competitive edge; most dynamic companies encourage the search for new and innovative approaches to manufacturing.

How does one go about investigating JIT further? Having got a firm grasp of the principles of the JIT philosophy, it is then helpful to test one's understanding by thinking of a particular problem and comparing the traditional approaches to the problem with those suggested by JIT. As an example, consider a forging process. In the traditional process, the raw material, a steel bar, is heated until nearly white hot and then forged into a complex shape. The part is then plunged into cold water to harden the outside. This results in frequent cracking of the part and a high failure rate in service resulting from hidden cracks. The JIT approach would look at solving the basic problem, in this case the rapid cooling. One possible solution, which has proved successful in practice, is to use air cooling after forging and more gentle heat treatment. This almost eliminated cracking and the in-service failure rate was very much reduced.

When the JIT philosophy has been mastered, the 'pathfinder' may then need to examine closely other JIT implementations (some examples are given in Chapter 9) after which the well-informed 'pathfinder' will be in a position to initiate his own implementation in a professional manner.

Not all companies appreciate the benefits that JIT may bring. The company may be successful and see no reason to change its methods. Alternatively, the company may be facing difficulties and its long-term survival may be in doubt; in which case JIT may be postponed until the situation improves. In practice, the second case is more common than the first, mainly because in a successful company the senior management are likely to be more relaxed and more willing to take a longer term view. If a company is facing severe difficulties the senior management are often extremely nervous and only thinking of the immediate future, sometimes as little as a month ahead. The irony, of course, is that the company in difficulty is the one most in need of JIT.

If a 'pathfinder' finds himself in a situation where it seems likely that the senior management will not respond well to the opportunities offered by JIT, he should not be discouraged, but wait, putting forward the case for JIT as and when the opportunity arises.

Even if it is anticipated that senior management may resist JIT, I have always found that once the background and the benefits of implementations are presented, possibly with the aid of an external consultant, then many senior managers can be persuaded to implement JIT. Indeed, by presenting the evidence in a straightforward manner, I have never had any trouble in persuading senior management that they ought to progress with JIT.

At the end of the basic understanding stage the 'pathfinders' have a good knowledge of JIT and what a good implementation entails; they are now ready to begin the preliminary education programme.

Preliminary education

The purpose of the preliminary education stage is to inform key personnel and top management about JIT. It should give them an overview of the philosophy, the steps necessary to implement JIT and the likely costs and benefits.

The personnel who participate in the preliminary education should be the key decision makers in the company – those who have the ultimate control over a project. These top decision makers normally include:

- General Manager
- Vice-President or Director of Manufacturing
- Vice-President or Director of Engineering
- Vice-President or Director of Finance
- Vice-President or Director of Marketing

as well as:

- Manufacturing Managers
- Sales Managers
- Purchasing Managers
- Engineering Managers

The exact number will obviously depend on the size of the company. In a small company, where functions may overlap the number of key personnel who need the preliminary education may be as low as five; in a large company, on the other hand, upwards of twenty-five staff may be involved.

It is vital that the preliminary education should be undertaken professionally and this usually requires the assistance and experience of a professional educationalist with substantial JIT experience.

I have always found that a preliminary education programme which is objective and does not overtly set out to persuade top management to undertake JIT is the most effective. Why? Because if JIT is oversold top management may become suspicious and evasive. The golden rule to follow in preliminary education is to be objective; present the evidence and let it speak for itself.

JIT is a major step and will therefore lead to fundamental changes in the way in which a company operates; this should not be glossed over during the preliminary education stage. On the contrary, everyone involved must be given a full understanding of JIT. One question often asked is how long should the preliminary education seminars last? If they are too short the necessary ground may not be covered, and if they are too long there is a risk that boredom will set in or that senior management will be reluctant to commit themselves to that length of time away from their desks.

When I give preliminary education seminars, I find that the best compromise is 16 hours of seminar *provided it is well structured and performed professionally*. These 16 hours can be split into two or three weekly sessions to allow further considerations of the implications. I have attended seminars which took 16 hours to say very little. Top management time is valuable and should not be wasted. Furthermore a poor preliminary education seminar is a poor advertisement for JIT.

Great care should therefore be taken that seminars are competently carried out. There is too much to lose if this stage is left to the inexperienced.

What should preliminary education cover? There are three basic questions that must be dealt with:

- What is JIT?
- What are the benefits?
- What will it cost?

Top management will want to know the answers to these questions before they can fully commit the company to JIT. If they are ready to look at JIT in more detail, the natural next stage is cost/benefit analysis.

Cost/benefit analysis

As I have said earlier, JIT can produce many benefits at relatively low cost. Companies rarely have a problem producing a financial justification for JIT. However, the cost/benefit analysis should be done with care for the following two reasons:

1. Consideration of only a small part of the benefits may indicate a lower rate of return than could be achieved in practice.
2. Concentrating on only a few benefits may give the wrong impression of JIT.

The costs/benefits associated with JIT can be divided into two types: the 'hard' costs/benefits, which are tangible and quantifiable, and the 'soft' costs/benefits which are more difficulty to quantify.

Hard costs/benefits	Soft costs/benefits
Inventory reduction	Increased sales
Work-in-progress reduction	Increased customer service
Increased productivity	Increased quality
Reduced obsolescence	
Reduced premium freight	

A good JIT implementation can achieve a significant reduction in inventory and work-in-progress levels; so substantial, in fact, that they justify most JIT implementations before the other factors are even considered. There is therefore the temptation to look at only one or two of the benefits to produce the final justification.

The rate of return from curtailed JIT justification may still be reasonable, but it is better to provide a breakdown of as many benefits of JIT as possible in order to show a much higher rate of return. This is worthwhile, for any decision to go ahead with JIT in the expectation of a low rate of return may result in a lack of full commitment on the part of top management. If more detailed figures are produced, showing a higher rate of return, they will demonstrate a more comprehensive justification for implementing JIT and gain a correspondingly full commitment by top management.

Another reason to avoid a narrow focus in cost/benefit analysis is that concentrating solely on, for example, inventory and work-in-progress reduction, may convey a wrong impression of JIT, implying that JIT only tackles inventory and work-in-progress control, whereas it is of course much broader.

Costs/benefits in terms of increased sales can be difficult to quantify. One company (see Chapter 9) doubled their sales when they implemented a JIT system, but an average sales increase of 30 per cent is more typical for most JIT implementations. This increase in sales comes about because products are more likely to be available when needed and to be of better quality. Manufacturing lead times are also shortened, which gives manufacturers in a make-to-order environment an advantage over their competitors.

The improvement in quality is particularly hard to quantify. It will certainly boost sales and this effect can be incorporated into the estimated increase in sales. Another effect may be to allow the company to increase the price charged for the product. Customers are usually willing to pay a higher price for high quality items, and this possibility should not be overlooked.

The productivity of both direct and indirect labour increases in a JIT implementation as less time is spent chasing up late orders, there is a lower risk of shortages and less rework.

The shorter manufacturing lead times combined with the smaller lot sizes mean that there is less manufacturing on the basis of anticipated future demand and less risk of a surplus once the demand has been met. This means that the risk of obsolescence is significantly reduced. There is also less risk of running short of components and consequent poor customer service. Depending on the industry, but particularly in cases where there is a high rate of engineering change, such as electronics for example, these benefits can be dramatic.

Many companies routinely have last-minute completion of orders or frequent shortages. In these circumstances they often resort to paying high freighting charges to ensure delivery. These rush deliveries can cost a significant amount of money when totalled over a year; better organisation and shortened manufacturing lead times resulting from JIT can reduce or eliminate high freighting costs.

58 Putting the Just-in-Time Philosophy into Practice

Let us look at a typical cost/benefit analysis for a typical company (Table 4.1). The figures given are for illustrative purposes only, and the benefits are for a good implementation. If costs, especially on education, are reduced then benefits will be lower.

Table 4.1

Annual sales	$95 million
Employees	950
Pre net profit	10% of sales
Annual direct labour costs	£9 million
Current inventories	$20 million
Annual purchase volume	$30 million

Costs	*Onetime*	*Recurring per year*
Process improvements including set-up time reduction inspection devices reliability improvements	$100,000	$20,000
Control improvements including pull type systems	35,000	15,000
Project team Full time project leader and assistant for 1 year	105,000	
Education		
Seminars/courses	160,000	20,000
Consultant	80,000	30,000
	$500,000	$85,000

[The process and control improvements are discussed in more detail in Chapters 6 and 7, the project team later in this chapter and education in chapter 5.]

Benefits

The benefits below (in terms of per cent improvement) are those that current successful implementations have achieved [see Chapter 9 and Sepehri (1986)]

	Current	Improvement (%)	Total
Inventory reduction	$20,000,000	50% at 18% (18% is assumed to be the carrying cost)	$1,800,000
Warranty cost reduction	400,000	40	160,000
Purchase cost reduction	30,000,000	5	1,500,000
Sales	95,000,000	30% at 10% (10% is the profit on sales)	2,850,000
Labour Productivity	9,000,000	7	630,000
Obsolescence	300,000	50	150,000
	Gross annual benefits		7,090,000
	Less annual costs		85,000
	Net annual benefits		7,005,000
	Net monthly benefits		583,750

Payback period
 one time cost/net monthly benefit = 0.86 months (that is, less than 1 month)
Return on investment (first year)
 net annual benefits/one time costs   1401%

It should be emphasised at this stage that the above figures are for a successful implementation: there are many pitfalls (Chapter 10) which may considerably reduce the benefits.

Some points about the figures can be highlighted. Allocation for education could be increased. Although an increase (even the doubling of the allocation) would not have much effect on the payback period or general JIT progress with the above figures, it could easily make the difference between a good and a mediocre implementation. It is also worth stressing that the net monthly benefit of $583,750 is the benefit expected from JIT every month after implementation, or, to put it another way, $583,750 is the amount lost for every month's delay in implementing JIT.

Commitment

Any successful implementation of JIT is greatly helped if top management are committed to it. Without such commitment, implementation can still prove successful, but it is obviously much easier if it is present.

Top management commitment is important for a number of reasons, the primary one being that of *authority* – top management have the authority to support the changes that JIT requires. Lower management will only be able to authorize investments up to a (often low) set figure. With JIT implementation, this may necessitate a whole series of authorizations to achieve the level of investment required. Each authorization often requires a detailed justification, the preparation of which can frequently divert the JIT team away from the more important work of the implementation itself. If top management is committed only one justification and authorization is required.

The second aspect of this commitment by top management and its consequent authority is in personnel management. For example, top management can appoint a high-quality employee to the position of JIT project team leader. Other personnel changes may become necessary in the course of the implementation and these may only materialize through top management commitment.

The third aspect is more psychological. If top management is committed to JIT and their commitment is *visible*, lower management are much more likely to follow. This commitment should be demonstrated and needs to be publicized throughout the company.

Once the top management of the company has made a commitment to JIT, I have often found it useful to put it into writing. This written commitment will not, of course, be a legal document but it will set out the implementation budget, the project team composition (the appointment of a full-time project leader with assistant should be announced although the individual will probably not be named at this stage), and the start and completion dates of the pilot plant implementation (again the actual pilot plant will probably not be identified at this stage). This written commitment saves later confusion about what is expected from the project team.

Go/no go decision

As part of the top management commitment stage the company must make the go/no go decision. If the company is not ready to deal with the changes or the financial investment involved in switching to JIT then they should wait. A no go decision does not mean that the company will *never* implement JIT, it may just mean that the implementation is delayed. Such a delay may prove to be tremendously costly to the company in terms of continuing high inventory investments, poor quality and low productivity, as well as competition from companies that are already successfully implementing JIT. The no go decision can therefore threaten not only the profitability of the company but may question their eventual survival.

The reason that this go/no go decision stage has been included is that a company *must* make a definite choice. A go decision should mean that the company will progress quickly and aggressively to JIT implementation. Any half-hearted approach should be avoided and will only result in an implementation that falls below expected levels of achievement while probably costing as much as a good implementation.

What are the factors that bear on a go/no go decision? A checklist can be made:

GO/NO GO CHECKLIST

Is the basic understanding stage completed?
Has the preliminary education been completed?
Does the cost/benefit analysis indicate a good return from JIT?
Is there top management commitment?
Are all relevant sections of the company involved?

Answering yes to every question on the checklist identifies a company that is ideally suited to JIT and that could rapidly proceed with implementation. However, such companies are probably in a minority. Many companies will not have fulfilled all the conditions listed above, and will be faced with an unclear picture as to whether they should proceed with JIT. The most likely questions to have negative responses are the last two which give an indication of the company commitment, particularly when the enthusiasm for JIT comes from only one section of the company. Those who are enthusiastic about JIT have two choices:

1. Put JIT on hold for a year or two, 'Plan A'
2. Proceed with a smaller scale implementation, which we can term 'Plan B'.

The second choice is the most suitable in the majority of cases. The smaller scale implementation may still offer a good return on investment and it will provide essential experience of JIT. When the time comes for a full-scale implementation, the company will be in an ideal position to gain from the implementation.

The decision to opt for a smaller scale implementation can arise from a number of causes including a nervous top management, lack of commitment from sales or purchasing, difficult market conditions, etc. The implementation should, nevertheless, be managed with the long-term view in mind and should therefore concentrate on a section of the operations that is small enough to demonstrate that even a limited implementation can make an impact. Later it can expand to include other sections of the company.

The go/no go decision stage is obviously the most crucial stage in JIT implementation. If the outcome is to go ahead with either a full or a small-scale implementation, then the next stage is to set up a JIT project team.

JIT project team selection

A high quality, dynamic project team is essential to complete a successful implementation of JIT. The majority of successful JIT implementations on which I have worked have had project teams with the following characteristics:

- a full-time project leader
- a membership of about 10 key personal
- an aggressive implementation schedule
- team members attending all the meetings
- meetings held once (or twice) a week for a maximum of two hours

We shall look at each of these points in turn. First the project leader should be full time. This will ensure the right amount of time and attention is spent on coordinating the project. For many small companies, a full-time project leader may seem an unnecessary expense, but this is a dangerous assumption and those companies who are tempted should ask themselves: 'How can we guarantee that the JIT implementation is successfully carried out *without* a full-time project leader?' The answer is that it may be possible but the outcome is by no means certain.

Other relevant points about the project leader; he should be:

- a relatively long-serving employee with good in-company experience
- a high quality manager – probably the best available
- a good communicator
- able to relate to people well – it will be necessary to motivate large sections of the company

JIT will form the basis of future company profitability and

survivability. A company implementing JIT should therefore select the best manager for the job of project leader. This may necessitate depriving another area of his skills, but JIT should take precedence over narrow, sectional interest.

However, the position of a JIT project leader is not easy. It is a demanding role and thus should be recognized, with the project leader taking a large part of the credit for a successful implementation.

With regard to other members of the project team, I have found that between 8 and 11 is the ideal number of members for most companies (although smaller companies may need fewer members). The project leader can manage this number of people on a one-to-one basis, and they can cover a broad spectrum of company activities. Some of the team may be the original 'pathfinders'. Let us look at an example of a JIT project team (Table 4.2).

Table 4.2 *Example JIT project team*

Project leader	Purchasing manager
Assistant leader	Manufacturing engineer
Members	Sales manager
	Manufacturing manager
	Plant manager
	Quality control manager
	Data processing manager
	Production control manager
	(Financial manager – part-time member)

This example, for a repetitive batch manufacturing company, gives the scope of the team. It is only an example however; in other companies the best project leader may well come from another area. The project team should meet once or twice a week for a maximum of one to two hours. I have always found that scheduling the meeting for late afternoon (from 4–6 p.m.) will often ensure that the meetings finish on time (or even early!).

It is important that all team members attend all meetings. It is fatally easy to slip into a pattern whereby only half the team are present for any one meeting, and subsequent meetings then spend a considerable amount of time reviewing the previous meeting, achieving little real progress as a result. If this happens too often schedules slip and deadlines are missed. One solution is to set up a mechanism to ensure full attendance; in most cases an attendance list for top management, who then check on absentees, is enough to keep attendance high. If the problem becomes acute, some companies have used the idea of 'boss default', whereby if someone cannot attend the meeting, their boss attends instead. This helps to ensure that absences from the project team meetings are kept to a minimum, and even if any absences are unavoidable, those who attend in the absentees' place can still make the necessary decisions.

When the team is fully convened, the first task is to determine an implementation schedule. For larger companies, this schedule should initially concentrate on a pilot plant to provide the experience of an implementation without the long lead time of full company involvement. In the case of a small company, of course, the pilot plant may well be the whole company. The implementation in the pilot plant should be aggressive, with realistic dates for achievements. I have found that an implementation schedule of one year for the pilot plant works well. This is short enough to maintain enthusiasm and ensure that changes in personnel do not affect progress too much. Remember, good managers will be sought after by other companies, and a longer time span increases the chances that key personnel may leave the company before implementation is complete.

Setting a longer time scale may seem attractive at first. For example, if we set the implementation time span to be two years, we have longer to carry out the individual tasks. But it also increases the risk of losing the focus of the project, reduced enthusiasm, and changes in key personnel.

A one-year schedule (see Chapter 9) will include the remaining four steps discussed in this book. One important point ought to be borne in mind at the outset: *JIT is a process of continual improvement*. We are not looking for one-off improvements that will produce results only in the short term, rather we are setting in place mechanisms that will result in continual improvements. This is illustrated with respect to stock turn in Fig. 4.1.

With JIT, performance ratings such as stock turn, productivity and quality should improve year by year after implementation. At the end of the implementation stage, the project team should be as much concerned with the *trend* of the performance ratings as with their absolute values. Any deviation in the performance ratings should be examined and measures taken to ensure that the improved trend is maintained.

This means that the task of the project team is not finished once the implementation is complete. Rather, JIT progress should be monitored after implementation and the project team should remain in place to maintain the upward trend in performance ratings. Further discussion of the implementation strategy is contained in Chapter 9 where two case studies are described.

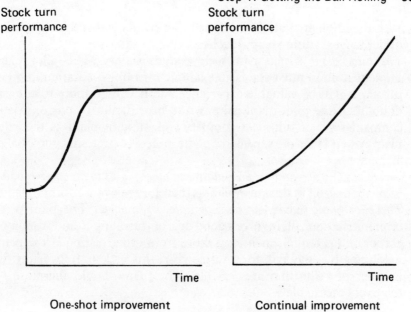

Figure 4.1 *One-shot and continual improvement*

Identifying the pilot plant

The final stage for the larger company in this first step involves identifying the pilot plant. This is carried out in parallel with selecting the project team since it is important that a significant element of the project team come from the pilot plant. As previously stressed, for large companies the best scheme involves selecting a pilot plant whose preliminary JIT implementation can be completed and then gradually introduced into the other plants, one at a time. The advantages of this approach are that the efforts of the project team are not spread over a number of plants simultaneously. Rather, the impact is focussed and results are obtained on a shorter time scale.

For a smaller company, the implementation can be accomplished across the whole company at a single stroke.

In this final stage the problem is to identify a suitable pilot plant. What characteristics should it have? Each company and each individual plant is different and so intuitive skills and judgement are important in assessing them. However, there are some general considerations which should be borne in mind.

- *The pilot plant should be relatively self-contained.* If there are many material and component flows to other plants, this could interfere

with the implementation since the other plants will still be operating under more traditional approaches.
- *The pilot plant should pose some real difficulties.* The pilot plant implementation must act as an example for implementation in other plants. It will be valuable experience and it will do more to convert doubters if the pilot plant is known to be difficult to control. Many companies select the most complex and difficult plant as the pilot plant; when JIT works in such a plant, it demonstrates its potential to all personnel.
- *The pilot plant should be representative of other plants in the company.* An implementation in this pilot plant is therefore relevant to other plants.
- *The pilot plant should not be geographically remote.* The pilot plant implementation will involve a good deal of travelling to and from it by corporate personnel, including those from other plants. If the plant involves a two-day trip for a single visit, this will severely affect the number of visits that are feasible. The plant should therefore be relatively accessible.

It is quite possible that no one plant meets all of the conditions, and some compromise may be necessary. This will be a matter of judgement but the important point to consider is that the pilot plant must provide the basis for a successful implementation. The degree of success of the pilot plant implementation may well determine the enthusiasm for and subsequent success of an implementation in other plants. It is therefore worthwhile spending some time on this stage.

Step 1: Conclusion

This chapter has described step 1 – *getting the ball rolling*. As stressed in the introduction, this first step establishes the foundation upon which the implementation can be built. The implementation of JIT requires a change in atmosphere and attitude within the company, and this first step will determine its direction. It is essential that it is done professionally and competently.

The time scale of step 1 will vary between companies, primarily because of the first two stages, basic understanding and preliminary education. These two stages may take either a few weeks or a few months to complete before the company is ready to start a cost/benefit analysis and lobbying for top management commitment.

An example of an ideal time scale for the whole of step 1 is shown in Fig. 4.2. This will vary from company to company, although step 1 should be completed within a minimum of four months in many companies. Extending the time scale can be beneficial where there are uncertainties that need to be clarified, but it can also be damaging, in that enthusiasm may wane.

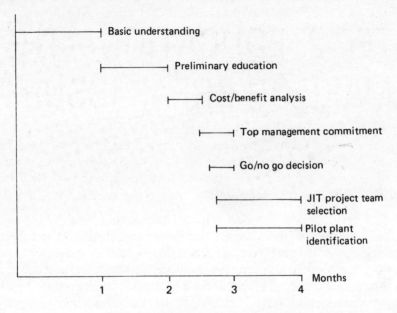

Figure 4.2 *An ideal time scale for step 1*

Once step 1 has been completed the company is then in an excellent position to achieve a good implementation by moving on to the remaining four steps.

5

STEP 2: EDUCATION – THE MAKE OR BREAK ISSUE

Introduction

This step involves educating all relevant personnel about JIT so that they fully appreciate the changes that will be required. It has been called the make or break issue because that is exactly what it is: if a company skims over this step the resulting implementation may well have many difficulties. A thorough education, on the other hand, prepares the way for the JIT philosophy to reach the whole company increasing the chances of a highly successful implementation with a continual improvement of company operations for years to come.

Education is therefore fundamental. The education programme will need considerable amount of thought before a viable and effective plan to educate the majority of the relevant personnel can be developed.

In one important aspect our task is made easy for although we must be careful not to oversell JIT, success stories demonstrate the advantages of JIT and show why the company should proceed with its implementation. This is a key topic when briefing top and middle management. Their personal success is closely bound up with that of the company. If the JIT implementation is successful, their security, career prospects and salary all increase. Using examples of other successful implementations can be a key factor in motivating top and middle management to implement JIT.

However, other aspects of JIT education are more difficult. We have to ensure that the JIT philosophy is fully assimilated by the company personnel to enable attitudes to change. It is this change in the way people do their jobs that distinguishes JIT from, for example, MRP and MRP II. Of course, MRP and MRP II both require changes in the way in which certain jobs are done. For example, the proper implementation of an MRP II system means the elimination of a previous, more informal 'hot' list on the shop floor. The production controller has to banish such 'hot' lists and must replace them with formal systems, since their continued use would weaken the operation of the MRP II system by overriding the priorities produced by the system.

The changes that accompany the introduction of JIT are different, in that instead of imposing greater regimentation within the company, we are aiming to alter the employees' perception of their roles and responsibilities. The company personnel need to consider which systems can be set up to identify problems (the responsibility of top and middle management), take corrective action once a problem is identified (the job of middle and lower management and direct labour) and to examine methods of eliminating waste (all employees). Instead of more regimentation, greater use of expertise and intelligence is required on the part of all people in the company.

This process of altering the way in which employees operate requires a much more comprehensive education programme than that associated with, for example, MRP II. This educational programme also needs to be ongoing. The JIT system is one of continual improvement; if implemented successfully we expect improvements for years to come. But this will only happen if we reinforce the message about JIT at regular intervals over the years after an implementation. In fact we can say that as long as we want the JIT system to mature and improve, we must still have an educational programme.

This chapter details the educational programme that must be developed to benefit fully from JIT. To cut back on education risks the future of the JIT system and possibly the future of the company. One example of this is a company in the southeastern United States that decided to set up a Kanban system (see Chapter 7) in one department. They did this without a proper educational programme. The result was low employee involvement and low levels of understanding, leading to missing Kanban cards and frequent stockouts. The Kanban system was eventually abandoned. Some time afterwards the company decided to implement a JIT system, this time with a proper educational programme that educated *everyone* who would be involved in the new JIT system. The result has been highly successful.

What is JIT education?

As I indicated in the introduction, JIT education differs from educating for other manufacturing management approaches in that some fundamental changes in attitude and behaviour are necessary. With previous approaches the aim in education was frequently to make more regimentation acceptable to the workforce – as long as they followed the formal system everything was fine. With JIT all employees have a greater role in helping the company to prosper. Achieving this fundamental change in attitudes and behaviour demands a comprehensive educational programme. Such a programme must be carefully thought out to make full use of the time available.

What should the educational programme consist of? Primarily it needs to achieve two aims.

1. First, it must impart an understanding of JIT philosophy and its application in industry.
2. Second, the programme should be structured so that the employees start to apply the JIT philosophy to their own particular job.

From my experience in giving seminars and running courses on JIT, I have found that if the first aspect is stressed and a thorough background to JIT is given, a good understanding of JIT results. Having achieved this, it is relatively straightforward to address the second aspect.

A deep understanding of JIT remains with employees for some time, although, of course, we do need to reinforce this from time to time.

We should not confuse this stage of education with training. Education involves giving the broader view, describing how each of the elements of JIT fit together. Training, by contrast, imparts detailed knowledge about a specific aspect. For example, an educational programme about Kanban systems would discuss how the Kanban system works, the role of the Kanban cards, which manufacturing systems it is suitable for, the potential problems, the advantages and disadvantages. By contrast, a Kanban training programme would, for example, tell a someone exacatly what to do when a Kanban card of a certain type is received. With a training programme there is no broader view. If someone using a Kanban system was given no education, but only training, he would not appreciate why he is doing what he had been trained to do. If this happens, the motivation to keep to the detailed training instructions may be lost, resulting in difficulties in the operation of the system. The distinction is profound: *education* gives the broad view, *training* gives a specific set of instructions.

Both education and training are necessary for some parts of a JIT system. Training can take place after education. I should stress again that education is necessary for *all* personnel. Without it there will not be a basic understanding of JIT and its effect on individual jobs.

Who needs JIT education?

This should really be rephrased: who *doesn't* need JIT education? *All* employees associated with the JIT system should go through a good education programme.

The programme will obviously differ for each group within the company in that, for example, top management will require a different approach to that used for the factory floor workers. The depth of coverage will also vary between groups. For example, there will be a group of key employees who will require a total immersion in JIT. This group will usually consist of the project team and all the manufacturing/production managers. By total immersion, we mean a complete, thorough educational programme that covers every activity in JIT.

Where should the education take place? The answer varies between groups. It is sometimes advisable to suggest that as many people as possible attend an external course in another city, but with only one or two from each company included in each course/seminar. External courses are advisable under certain circumstances:

- *For basic understanding*, the pathfinders find it convenient to attend an external course/seminar.
- *For in depth coverage*. Where one aspect of JIT needs to to be covered in depth, the small number of relevant personnel can attend an external course.
- *For certain top management*. Some top managers feel uneasy attending an on-site course/seminar. This may be because a manager may feel less informed than he/she would like to appear. Having attended an external course/seminar they are subsequently happy to attend in-company seminars.

However, I find that in-company courses have a number of advantages:

- People who are going to have to work together implementing JIT, attend the course and discuss potential problems together.
- The course can be more closely structured to reflect the company's interests.
- An in-company course helps to remove barriers within the company.
- Less travelling is involved.
- In-company courses are usually less expensive per employee since there are no travel or hotel costs.

There are some potential problems associated with in-company courses, most of which are avoidable:

1. *Constant interruption.* Some managers feel they cannot be out of reach of their subordinates and arrange to receive messages about crises that occur and may leave the course temporarily to attend to these problems. This causes, first, an interruption in the course and, secondly, the manager to miss some course material. I avoid this problem when I run courses by holding the courses off-site, usually in a local hotel, and by insisting on no interruptions, messages and absences.
2. *Conflicts within the company.* These can appear in the form 'Fred in sales has been messing me about for years and now I can take this opportunity to get back at him.' In discussions during the course, Fred is verbally attacked, he responds with a counter attack, and in the resulting antagonism the relaxed format necessary for free discussion evaporates. I find the solution is to cut short such attacks as early as possible.
3. *Attendees drawn from wide areas.* When courses are run in-company the personnel attending each course (numbering roughly 20) will probably cover a wide area of activities within the company. Some may come from purchasing, others from production, and so on. This means that some time has to be spent in discussing each aspect of JIT from a variety of points of view. Some would view this as a disadvantage, but I have found it to be a tremendous benefit. It offers an opportunity to break down departmental barriers and to allow each person to understand the problems of others. In one course I was conducting, the production planners had a history of continual complaints about shortages which they blamed on purchasing. During the course, one person from purchasing explained at length the problems that they had, which were mainly associated with poor suppliers. This gave the opportunity for the production planners to understand purchasing problems more fully and from there on they worked very closely together with purchasing in successfully implementing a JIT system.

Most companies will have a mixture of in-company and out-of-company seminars and courses in their education programme. Each has advantages, and both can enhance the quality of the educational programme.

What should be covered?

Since education is a key part of a successful JIT implementation, it should be well thought out. The educational programme should be designed with the assistance of a JIT educational professional who should help to structure the programme to reach all employees affected. This professional should be chosen with care; he should be highly experienced, both as an educationalist and in implementing JIT.

Having selected an educationalist the process of designing the educational programme commences. One aspect which ought to be stressed again is that we are not talking about a one-off JIT educational programme: the JIT system should continue to improve for years to come. Any educational programme must therefore be divided into parts:

1. Initial education to cover the implementation phase of JIT.
2. On-going education to cover the period after the JIT system has been implemented.

The second part of the educational programme can be more low key than the first, but it cannot be overlooked without jeopardizing the long-term improvements. Each relevant employee should probably attend a formal session every two months or so, but will have a more informal discussion session every other week. The emphasis on this on-going education should be, first, to reinforce JIT philosophy and second to encourage discussion during which problems can be brought out into the open and solutions found. These more informal sessions can be based around those in a small section of the company along the lines of the productivity circles described later in this book. If problems still remain unsolved they can be discussed at more formal sessions. I have found it very useful at these informal sessions, if problems are not only discussed but also some of the solutions that the groups have found are publicized. The point is not only that this gives the group some credit but also that the solutions can often be of use to other groups who may be able to apply them to their own problems. Later in the book an example is given of one group in a company who decided that they could operate with three people in a group instead of four; this was adopted throughout the company resulting in a productivity increase of 25 per cent.

As far as the initial education programme, which is of most immediate concern to those in charge of the implementation, is concerned, the major points are these:

- The education programme should be structured to reflect the levels of management. Top management require a different programme from middle management, who again require a different programme from lower management and so on.
- Small groups in the company should have an enhanced educational programme which results in their being totally immersed in JIT. This key group, which includes the project team, can then act as a reservoir of knowledge and expertise for the rest of company.
- Education should continue after the implementation to ensure that JIT is adhered to.
- Everyone who is going to come into contact with the JIT system should be educated. For most JIT implementations, that means everyone in the company.
- The education programme should not only impart knowledge about JIT but it should also initiate discussion and thought that results in a behaviourial change in the employees. JIT requires, for most people, a fundamental change in behaviour, which will only come about when the impact of JIT is fully assimilated.
- Most education programmes will consist of both in-company and out-of-company courses and seminars.

The initial educational programme prior to implementation needs to stress the impact of JIT on the operation of the company and on individuals within the company. JIT needs to be assimilated by the employees and this can be achieved in two ways. First the educational programme should allow sufficient time for discussion about the impact of JIT on job functions. Second the educational programme should have time for sufficient repetition of the essential points. After the implementation, education will continue and the assimilation of the JIT approach will be further reinforced by subsequent tuition.

One of the major aims of the education programme should be to arouse the enthusiasm of the workforce. This will greatly aid the implementation. Educators should not fall into the trap of putting across the message that 'everything about JIT is wonderful', promising a wealth of benefits with little effort. I always find this to be counterproductive, for when JIT is eventually implemented and a problem is encountered the work force will be unprepared for problems and confidence in JIT will be lost. In fact, when several problems occur together confidence can be reduced to zero.

Instead I present an objective view. This can be done by stressing that JIT can produce many benefits but that to achieve a good JIT implementation requires, first, hard work; second, a change in attitudes and work practices and, third, a willingness to face up to problems and solve them in a cooperative spirit. Those attending the course maintain a high degree of enthusiasm because they have learned from the course what JIT can achieve but they are also prepared to work at problems and solve them in order to end up with a successful JIT implementation.

Possibly the most important message that needs to be demonstrated to the workforce during the JIT educational programme is that the most important people in the JIT implementation are not top management or even the project team but the workforce. Top management can give the go ahead, the project team can organize the implementation, but JIT will not work unless the factory floor staff wants it to.

In one company where I was conducting a course on JIT, the workforce were extremely negative towards JIT at the start. Every six months the company had a campaign on one issue or another, such as zero inventory or zero defects, and the workforce was convinced that JIT was just another campaign. For these previous campaigns, the management had not bothered with a structured educational programme and the JIT course was the first course that the vast majority of employees had attended. It took two days for their enthusiasm to surface and for cynicism to be swept away. After the course the workforce were 100 per cent behind JIT and, what is more important, they were realistic about the problems and difficulties ahead. This company went on to complete a very successful JIT implementation.

Step 2: Conclusion

Step 2 – Education is the make or break issue, and if carried out properly it will give tremendous momentum to a JIT implementation.

This chapter has stressed a number of points about education:

- Education is fundamental; it is not just an option.
- The educational programme needs to be more comprehensive than is the case with, for example, MRP or MRP II. JIT requires a fundamental change in attitudes.
- The educational programme needs to be on-going even after the JIT implementation has been completed. Continual reinforcement of JIT helps considerably in attaining the goal of continual improvement.
- Training in JIT can be given after initial education has been completed.
- Most education programme will include in-company and out-of-company courses.

- All employees associated with the JIT system should go through a thorough educational programme.
- The education should be relevant to those attending the course. Top management will require a different programme from factory floor personnel.
- The educational programme should encourage employees to think about how JIT will affect their job.
- Repetition helps retention of information.
- Emphasize benefits but also discuss potential problems.
- The use of a high-quality professional educationalist will help substantially in developing an effective educational programme.

The education programme should be well thought out. We can truly say that success of an implementation depends to a large degree on the quality of the education programme.

6
STEP 3: PROCESS IMPROVEMENTS

Introduction

The first two steps aim to provide the fertile environment that is so important for successful JIT implementation. This chapter describes the third step, which is concerned with physical changes to the manufacturing process that will improve the flow of work. This is an essential feature of JIT: if the manufacturing process is not changed then it can become extremely difficult, if not impossible, to achieve JIT production.

The process changes take three main forms:

1. Reducing set-up time.
2. Preventative maintenance
3. Changing to flow lines.

Set-up time is the time taken to change a machine so that it can process another type of product. In the majority of western manufacturing industry there has, to date, been little attention paid to reducing set-up time. Excessive set-up time is harmful for two main reasons. First, it is time when the machine is usually not producing anything; thus long set-up times will reduce the efficiency of the machine. Second, the longer the set-up time the larger the batch size will tend to be for, with a long set-up time, it is not economical to produce small batches. With large batches come the disadvantages of increased lead times and increased inventory levels.

Reducing set-up times will therefore increase machine efficiency, decrease batch sizes, decrease lead times and decrease inventory levels. Another effect should also be mentioned: many companies have found that by attacking set-up times the set-up becomes more uniform and the scrap often associated with set-up is reduced. One company in the automobile component area, for example, found that the set-up time reduction programme decreased set-up scrap by 90 per cent.

As the inventory levels are reduced in a JIT implementation, unreliable machines become more of a problem. The reduction in buffer stocks means that if a machine should breakdown, subsequent machines quickly become starved of work. In order to prevent this happening, the JIT implementation will have to include a *preventative maintenance* programme to help ensure high process reliability. In keeping with other aspects of JIT, this can be done by giving operators responsibility for routine maintenance. This chapter details how a preventative maintenace programme can be implemented.

The flow of work through the manufacturing system can be helped by replacing the more traditional process layout with flow lines (Chapter 3). Using flow lines, work can flow quickly from process to process since the following process will be adjacent. In this way lead times can be significantly reduced.

Overall, the process changes described in this chapter can form a major part of the implementation of JIT within a company. The other steps also have to be accomplished, but these improvements, together with step 4, form the bulk of the actual implementation of JIT.

Set-up time reduction

A set-up operation may involve substituting one tool for another on the machine, replacement of a jig/fixture or resetting the machine for a new product. Sometimes the set-up time required will vary with the new product. For example, a paint sprayer can often be changed over from a light to a dark colour paint with little set-up time; the dark paint merely runs through the sprayer for a few seconds to remove any of the light colour remaining. However, when changing from a dark colour to a light colour the set-up time required to flush the dark paint from the sprayer can be much longer, since even a small amount of dark colour remaining in the sprayer can produce dark spots on the light coloured surface. Set-up times that depend to some extent on the sequence of products are termed sequence-dependent set-up times and are an important consideration in most manufacturing companies.

Set-up times can be considerable: it is not unusual to find that it takes eight hours to set-up a machine to run for perhaps one hour. This is a clear instance of the kind of waste which JIT aims to eliminate. Long set-up times are wasteful not only because they reduce efficiency and increase batch sizes, but also because they:

- increase inventory levels
- increase the risk of obsolescence
- reduce flexibility

Set-up time lowers efficiency in that when a machine is being set-up it is not producing anything. For example, if in a 40-hour week of operation a machine is being set up for 10 hours then it is only available for useful work for 30 hours; its maximum efficiency is therefore 75 per cent. If the set-up time is reduced to 1 hour then its maximum efficiency is raised because it is now available for useful work for 39 out of 40 hours. As well as machine efficiency, human productivity also suffers with excessive set-up time. An operator usually has to be available to carry out the set-up operation which may also require the services of a skilled set-up technician.

If a set-up time is long it only makes sense to do a set-up operation if there is enough work to justify that time. The consequence of this is pressure to increase the lot (or batch) size. This tendency has been institutionalized in the use of EOQ-type formulae for determining the lot size (see Chapter 3). The set-up time is an important factor in these formulae, and can mislead managers who believe that by using the formula to determine the lot size they are obtaining the 'optimum' lot size whereas, in fact, the lot size obtained is only the optimum given the assumptions behind the formula, which include a fixed set-up time.

Companies in south east Asia have shown that the set-up times are *not* fixed; they can be reduced. Sugimori *et al.* (1977) give the figures shown in Table 6.1 when comparing press plant performance between Toyota in Japan and representative automobile manufacturers in other countries. As can be seen from from Table 6.1, the set-up times in the Toyota plant are substantially lower. Such comparisons have recently led automobile companies in the other countries to put considerable effort into reducing their own set-up times.

Table 6.1 *Set-up comparison between automobile manufacturers (Sugimori et al., 1977)*

	Toyota	A (USA)	B (Sweden)	C (W. Germany)
Setup time (hours)	0.2	6	4	4
Number of set-ups per day	3	1	–	0.5
Lot size (days usage)	1	10	31	–

The large lots caused by excessive set-up time are not usually used immediately which causes another unwanted side effect; that of increasing inventory levels. This, in turn, can lead to an increased risk of obsolescence. For example, a company in England which produced in lots of about six months usage found that a large portion of its product line had been made obsolete almost overnight by a new technological development. As a result the company suffered a large loss when it had to write off a large section of its inventory. If the company had produced in lots of, say, one week's usage its loss would have been far less.

In combination, all these factors result in yet another drawback associated with excessive set-up times: reduced flexibility. As the market environment changes the manufacturer with excessive set-up time and large lot sizes is increasingly at a disadvantage. The company with short set-up times is much more flexible in its ability to respond to change. With increasing world competitiveness in manufacturing industries, increased flexibility can mean the future viability of a company.

How is low set-up time achieved?

Toyota in Japan have, over the past twenty years, made a systematic attempt to reduce set-up time in their operations. They have achieved remarkable reductions leading to what they call single set-up, meaning set-up times of less than ten minutes and frequently of under one minute (which they term one-touch set-up). As with other aspects of JIT, the reduction of set-up time in Toyota is regarded as an area of continual improvement, with the aim being to achieve one-touch set-up for all operations.

In a similar manner, a successful JIT implementation will usually include emphasis on further reduction of set-up times both during and after the initial implementation. This reduction can be achieved by the following:

- separate the internal from the external set-up
- convert as much as possible of the internal to the external set-up
- eliminate the adjustment process
- abolish the set-up step itself

Separating the internal set-up from the external set-up involves identifying those areas of the set-up that require the machine to be stopped (the internal set-up) and those that allow the machine to be kept running (the external set-up). For example, adjusting a fixture may be done with the machine running whilst placing a fixture on the machine may require that the machine is stopped. These two elements, the internal and external set-up, should be rigorously identified and separated. The operator can only work on the internal set-up when the machine is stopped, never on the external set-up.

Converting as much as possible of the internal set-up to external set-up involves performing as much as possible of the set-up when the machine is operational. For example, fixtures can be prepared, tools can be sharpened and adjusted, etc, before stopping the machine. By moving as much of the set-up operation as possible to an 'off-line' mode the machines can be kept running for the maximum amount of time.

Frequently quoted figures suggest that adjustments occupy around 60 per cent of the total set-up time. Consequently by eliminating the need for these adjustments the set-up time can be considerably reduced. The basic principle behind this concept is to change a continuously variable adjustment into a small number of discrete steps. For example, the range of adjustments to the end stop on a drilling machine can be reduced to a finite set by the use of steps on the shaft. Toyota have made wide use of such methods in moving towards one-touch set-up.

Abolishing the set-up step itself is the final set-up reduction concept. This can involve two aspects. The first is to standardize the parts so that the product range is reduced. Each part may then be used on a wide variety of products, thus reducing the set-up time problem. The second is to make the required parts at the same time either on the same machine or on parallel machines. For example, if two parts which are always assembled together for the final product are both made on the same type of lathe, they can be made either together in one operation and then split up or they can be made simultaneously on different lathes. This will not only reduce the set-up time but will also help to ensure that parts are made in matched sets.

Several techniques can be used to implement these four concepts:

1. Using quick fasteners can significantly reduce set-up time compared to more standard fasteners such as nuts and bolts.
2. Using a mechanical aid such as a ram can reduce set-up time, especially for heavier fixtures.
3. Standardizing the set-up as much as possible will help to ensure that the set-up operation becomes routine. Of course, it is probably only cost effective to do this for part of the operation.

4. Arranging the set-up so that it can be carried out by two people simultaneously will reduce the internal set-up and also reduce the time the machine is not working.
5. Attaching fixtures to standardized plates will allow these standardized plates to be quickly mounted on the machine when the machine is stopped.

As in all process operations, attention to detail is vital. For example, in one company (company A in Chapter 9), video tapes were made of each set-up and studied in detail. One operation involved tightening four bolts in order to attach a fixture to a machine. The video tape showed that each bolt required 13 turns before it was fully tightened. After closer examination, it was discovered that only the last two or three turns of each bolt actually tightened the fixture. The first step taken was to saw off each bolt to reduce the bolt tightening by two or three turns. Eventually quick release fasteners were substituted, reducing the set-up time even further. This example also illustrates one further point about set-up time reduction, which is that it usually involves little capital expenditure. Simple improvements can often produce significant reductions. This illustrates the point that the implementation of JIT essentially relies on using inexpensive methods to improve performance.

The reconfiguration of the factory floor, discussed in Chapter 3 and later in this chapter, is also a powerful tool in set-up time reduction. The traditional factory, as described in Chapter 3, is laid out in accordance with manufacturing processes, so that, for example, lathes occupy one section, milling machines another, grinding another and so forth. This results in inefficiencies through increased transportation. By contrast, a JIT factory is usually laid out according to product: flow lines are arranged for each product family. This helps to reduce transport difficulties. One further advantage of the flow line layout is that each flow line will handle a relatively narrow product range and will therefore have limited set-up requirements. One would therefore expect that the flow line-oriented layout associated with JIT systems would have considerably less set-up time compared with its traditional counterpart.

Preventative maintenance

Cutting set-up time reduces the amount of time that a machine is not running; to reduce the amount of down-time even further the number of breakdowns must be cut. JIT systems rely on preventative maintenance programmes with the aim of preventing breakdowns rather than repairing them once they occur.

The fact that a successful JIT implementation will reduce inventory and work-in-progress to a minimum means that the manufacturing system becomes more vulnerable to breakdowns. For once a machine does break down there are only small amounts of buffer stocks available and so subsequent machines rapidly become starved of work; whereas without JIT there is more time to repair the machine before the buffer stock is exhausted.

The treatment of machine breakdown in JIT is a classic example of the 'river of inventory' (see Chapter 3). As the inventory and work-in-progress levels (the level of the river) are reduced problems caused by unreliable machines are encountered (rocks are exposed). The traditional approach has been to quickly cover over the rocks to hide the problem. With JIT, we are now more concerned with solving some of the fundamental problems (removing the rocks) which can be done by setting up a preventative maintenance programme to improve the reliability of the machines. This is often referred to as total preventative maintenance (TPM).

In a conventional manufacturing system, therefore, breakdowns are not particularly important restrictions to the flow of work. There are usually large buffer stocks to help ensure that other machines are not starved of work in the event of a breakdown. In these situations the only machines that are affected by breakdowns are bottleneck machines. These are the machines that are running at full capacity and a breakdown means that some production will be lost, whereas for non-bottleneck machines breakdown time can often be easily made up. However, with a JIT system, the buffer stocks have been so reduced that all machines are in some sense bottleneck machines and a breakdown will reduce effective utilization of equipment, and hence lower efficiency as well as increasing shortages and increasing lead times. That is, breakdown will remove some productive time from the machine and thereby lower both the effective utilization and efficiency. Since there is little buffer stock, shortages result and the overall effect is an increase in manufacturing lead times.

Unreliable machines are therefore in direct conflict with JIT and reliability must be improved through the use of a TPM programme. Within a TPM programme one principle guides progress; to decentralize maintenance as far as possible to the operators. Preventative maintenance undertaken by the operators will include the relatively straightforward but essential tasks of:

- checking lubricant levels in the machine
- adding lubricant where necessary
- checking for wear
- noticing and acting on any unusual noises, vibrations, etc.

More major maintenance tasks can still be undertaken by the maintenance department. Devolving responsibility for the relatively straightforward aspects of maintenance to the machine operators has two major advantages which will increase the effectiveness of the programme. First, the operators are probably the workers who will know most about the operation of their machines and will therefore be the best people to detect unusual noises, wear or vibration. Second, the arrangement gives the operators a sense of ownership of their machines and they feel more responsible for ensuring that their machines avoid breakdowns.

Such decentralization also results in a better maintenance job. For example, when the operators took over routine maintenance at Harley Davidson during the implementation of their JIT system they found that some bearings on the machines had never been lubricated.

In order to achieve a high level of quality in the TPM programme with routine maintenance decentralized to the operators, there must be a comprehensive education programme. This programme, included in step 2 (Chapter 5), usually involves the maintenance department training operators in maintenance procedures. Some companies have then monitored the operators to ensure compliance but soon found that, with the right emphasis in the education programme, this was unnecessary and the operators did an excellent job of routine maintenance.

When should preventative maintenance be done? The answer is that it depends upon the circumstances. Routine maintenance can usually be accommodated in the normal production runs when, for example, no Kanbans (see Chapter 7) have been received by an operator. Such lulls are to be expected in normal operation. But time will also have to be found for more major maintenance operations. For most companies this is relatively straightforward; if there is single- or double-shift working major maintenance can be done in the non-production shifts. Three-shift operations, however, do pose more of a problem since there is no non-production time. The practice of three-shift operation is still prevalent in western companies, but Japanese companies prefer instead to have one- or two-shift operations to allow for TPM outside normal production hours. When there is three-shift operation, some non-production time has to be made available. Weekends or holidays are obvious possibilities but this can mean that some maintenance becomes overdue before the next weekend or holiday. It is therefore important to have some mechanism to allow for non-production time, and the best way may be to switch to two-shift working to raise overall efficiency. The argument for this is there is little point in having production 24 hours a day if a machine is down for eight hours or, worse still, if it produces poor quality parts or scrap for eight hours. It is far better to allow some non-production time for TPM so that the time spent in production results in high quality items with little machine down-time.

High machine reliability is therefore central to effective JIT in that it allows lowering of buffer stocks and an improvement in overall system performance. It can also have other benefits by increasing shop floor morale with the operators gaining some feeling of ownership over their machines.

Changing to flow lines

As we saw in Chapter 3 one of the principles of JIT is the constant effort to strive for simplicity. As already indicated, one way in which this can be realized is by rearranging the factory floor away from the process layout (Fig. 3.4) towards product layout using flow lines (Fig. 3.5). This will greatly simplify the management of the factory. Instead of an extremely complex operation based on a process layout where each product follows a lengthy route through the factory, we now have a simple, unidirectional flow with a limited range of products being produced on each flow line. For example, a Westinghouse plant in Asheville, North Carolina, changed their conventional process layout to five flow lines (which they call mini-factories) with a manager in charge of each one. The managers of each flow line found that they could control it much more easily than the original process layout and the move to flow lines helped to reduce manufacturing lead times by around 65 per cent.

As part of the switch to product flow lines, many Japanese companies now employ a flexible workforce with each worker having skills in a variety of areas. This is called *shojinka*. The use of *shojinka* allows the manning level of each flow line to rise and fall with demand for a product family. When demand is high each machine has one or more workers to operate it. When low, each worker may have two or more machines to operate. If there is no demand for a particular product family then all the workers are reallocated to another product flow line. This can, of course, only be effective if the workers are trained to operate several different machines and there is a suitable employee and union environment. In some industries such flexibility may take years to achieve.

At present only a few western companies have made any significant progress in adopting the concept of *shojinka*. There are a number of reasons for this, including excessive division of labour skills into a large number of labour grades. To achieve increased flexibility among the labour force requires the removal of most of these divisions and the creation of incentives for workers to be trained in a variety of different operations.

Not all flow line layouts are equally effective in helping the smooth flow of work through the factory and the maximum use of flexible labour. The birdcage layout shown in fig 6.1 restricts the use of flexible labour since there are few options for restructuring the use of workers. It also makes synchronization between workstations difficult, thereby increasing work-in-progress levels.

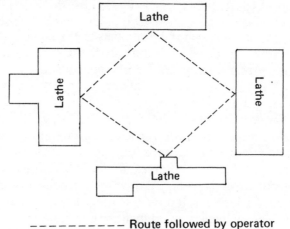

– – – – – – – – Route followed by operator

Figure 6.1 *Example of a birdcage layout*

Another layout which is undesirable is the isolated island layout, shown in Fig 6.2, which is an example of a product flow line being so small that it becomes difficult to adjust the number of workers and so reduces flexibility.

The ideal layout is often the U-shaped flow line dedicated to a particular product family. This layout simplifies control, allowing gradual reduction of inventory and work-in-progress levels. The advantages of the U-shape over a linear flow line are, first, that it assists communication, since workers on a particular flow line are physically closer to each other. The operator, for example, of the last machine in the flow line can easily tell the operator of the first machine of a quality

Step 3: Process Improvements 87

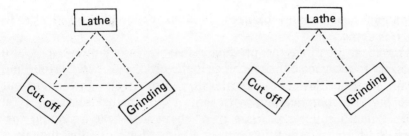

Figure 6.2 *Example of an isolated island layout*

problem arising from the first operation and action can be taken quickly. The second reason that U-shaped flow lines are preferable to linear flow lines is that the layout allows the workers access to a number of machines (see Fig. 6.3), each worker being physically closer to more machines than in a linear flow line and therefore able to operate several machines.

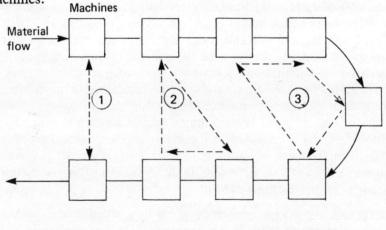

Figure 6.3 *Example of the flexibility of workers in U-shaped lines*

In order to establish suitable flow lines the first step is to divide the product range into product families. In many cases this will be relatively easy – each family will be one distinctive part of the product range. In other cases, when there is a finer distinction between products, the division can be more difficult and some sort of group technology analysis may be needed, often using coding and classification systems.

As well as dividing the product range up into families it is also necessary to reorganise the machines in the process-oriented layout into suitable product flow lines. This again is usually straightforward but there may be complications with large facilities that service several product families. Examples are paint shops and heat treatment furnaces. Splitting such facilities into small units for individual flow lines is not usually possible; they may have to be shared amongst several flow lines.

Associated with the move to flow lines is the gradual improvement in operations to improve efficiency. *Gradual* improvement is central to the JIT philosophy and the aim is to allow improvement in operations without large-scale disruption. JIT also stresses the need to improve human operations first before investing in automation. This approach is followed for a number of reasons.

1. The improvement in manual operations will avoid disrupting the operation of the company.
2. The cost of improving human operations is usually low compared with investing in automation.
3. Changes in manual operations can usually be reversed easily if they are not successful, whereas the investment in automation is often wasted if the automation does not work as anticipated.
4. JIT stresses the idea that we should be concerned with improving the overall system performance. To do this we initially improve the overall flow of work through improvement in human operations. Only when the overall flow in the manufacturing system is satisfactory is investment in automation justified. Putting automation investment ahead of investing in human operations is therefore putting the cart in front of the horse.

Top priority should therefore be given to gradual improvement in human operations to achieve good overall system performance. This can be aided by an effective TPM programme which stresses reliance on the present equipment, with modifications where necessary, rather than a large investment in sophisticated automation. JIT also means emphasizing investment in low cost machines that are flexible in their response to changes in product volume and product type. There is little point in investing heavily in specialized machinery if demand falls for those products that the machinery is designed to produce. JIT implementations are characterized by flexible machines that can be quickly altered from one product to another.

Conclusion

This chapter has described step 3: process improvements, which involve changing the in-house, physical aspects of production so as to help achieve Just In Time production. This includes set-up time reduction, preventative maintenance and moving to flow lines. Improvements in each of these areas will aid the smooth and efficient flow of work for more simplified and effective management.

Excessive set-up time has a number of drawbacks which hinder JIT production. To date little attention has been paid to reducing set-up time in western countries despite research which indicates that these are high in western companies when compared with industries in south east Asia.

Set-up time can be significantly reduced by a systematic programme which examines each excessively long set-up time in turn. Central to this are the four principles described in this chapter:

1. separate the internal set-up from the external set-up
2. convert as much as possible of the internal set-up to the external set-up
3. eliminate the adjustment process
4. abolish the set-up step itself

These four concepts form the foundation of set-up time reduction programmes which also emphasize attention to detail and involvement of the operators, since it is they who know most about a particular operation.

Further reductions in set-up time can usually be achieved by the move to flow lines where each flow line manufactures a more limited range of products. Other advantages of moving to flow lines, usually in a U shape, are that it eases the management problem as the lines are relatively simple and quality is improved as operations are closer together.

The other aspect of process improvements for JIT production is preventative maintenance. Unreliable machines cause both quality problems and increases in lead times and thus increased work-in-progress levels. A preventative maintenance programme will involve shifting maintenance activities to the operators so that they become responsible for routine work, with the maintenance department retaining responsibility for more major repairs. This helps promote the feeling of ownership of the machines by the operators and can result in a more effective job being done since the operators can frequently deal with problems before they become serious.

It should be stressed that each of the activities to improve manufacturing processes outlined in this chapter are of low cost. Large capital purchases are seldom required for a successful JIT implementation, instead a reorientation of the workforce is the major requirement. These process changes allow the full benefits of control improvements (Chapter 7) and vendor/customer links (Chapter 8) to be implemented.

7
STEP 4: CONTROL IMPROVEMENTS

Introduction

The way in which the manufacturing system is controlled will determine the overall results of the JIT implementation. The control policies adopted (along with process improvements and vendor/customer links) will determine the work-in-progress levels, the manufacturing lead times and the customer service levels. Improving the manufacturing control is therefore an integral part of the JIT implementation.

The principle of striving for simplicity provides the foundation on which the effort to improve the manufacturing control mechanism is based. Instead of complex control of a complex manufacturing system, JIT emphasises the need for simple control of a simple manufacturing system.

This simplified control revolves about a pull-type system and the use of quality control at source (using the techniques of statistical quality control). Instead of pushing work into the factory the pull-type systems pull work through the factory so that a job only progresses to the next workstation if there is adequate capacity. Be warned: conceptually the push-type systems do seem to have advantages in that they allow for the detailed planning of future events but, as with a lot of other aspects of manufacturing control such reasoning can be deceptive. For actual implementation, the advantages that pull-type systems offer are:

- simplicity, with no necessity for complex computer control
- once implemented, pull-type systems can be gradually improved to reach extremely high levels of efficiency.

This is not to say that pull-type systems are suitable in every instance. Where some long delays are inevitable (for example in the case of certain raw material or component supplies) a more plan-oriented approach such as MRP or MRP II may be useful. But the aim should always be to gradually reduce these long lead times wherever possible. This is discussed in more detail in Chapter 8.

Many of the changes described in Chapter 6 have increased the probability of successful control. In particular the move to small flow lines, the reduction in set-up times and improvements in machine reliability mean that many of the constraints on control have been removed. The small flow lines can act as semi-automated factories and are much simpler to control than larger, more complex factories. The reduction in set-up time means that there is no minimum lot size. Machines can now switch from one product to another more easily, thereby eliminating the need to plan the complex interactions between set-up time and priorities. Machines are also more likely to be available when needed.

With many of the complicated factors removed we can now superimpose a simplified control on this simplified factory. This chapter describes the essential ingredients of this simplified control. The pull-type system is outlined and the manner in which control is moved from centralized to local control is detailed. This local control includes quality control (statistical quality control) and the use of pull/Kanban systems. Finally, links between the JIT factory and the coordinating MRP (or MRP II) system are discussed.

Once step 4, control improvements, has been successfully carried out most of the in-house JIT system is in place. As with all the other steps, step 4 should be carried out with much thought and pre-planning, preferably with the aid of an external consultant.

Simple control

Simple approaches offer a number of advantages over and above ease of implementation. When they are used the whole control operation becomes clearer to the user. With a more complex control system such as MRP II, a supervisor who takes a job out of its priority list and completes it before other higher priority jobs may not know the effect of his action. In a simple control environment, such as a pull/Kanban system, the supervisor will readily see the effect, which is usually the subsequent operator running short of work. The supervisor can therefore make his own decisions more efficiently and effectively.

A simple system is also likely to be more robust. If there is even a small failure in the computer memory, disk drive or on data access then the output of a MRP system will be suspect. In contrast, properly designed pull-type systems are very robust and only a major occurrence

Step 4: Control Improvements 93

will significantly disrupt the operation of the pull/Kanban type system.

With these advantages why is it that it has taken so long for these pull-type systems to be implemented? First, responsibilities in western companies have traditionally been compartmentalized. Production control has usually been entirely separate from manufacturing engineering. Pull-type systems rely on many changes (outlined in Chapter 6) to be fully effective. Process improvements have traditionally been the responsibility of manufacturing engineering, which very rarely works closely with production control to bring about the changes needed. The result has been that production control has accepted the complex factory as a fact and has then proceeded to establish complex control systems (MRP, MRP II, or OPT) to control them.

The second and more problematical reason has been faith in technology. There has been a widespread belief that computer technology (both hardware and software) can be refined until it becomes powerful enough to give real-time control of the factory. If we are faced with driving a blunt nail into a hard surface we can take two approaches. We can get a bigger hammer until the task is accomplished or we can examine the nail, sharpen it, and then use a very small hammer. The use of increasingly sophisticated computer technology is equivalent to taking a bigger hammer. One result has been the poor implementation of MRP systems. In one survey [Anderson *et al* (1982)] only 9.5 per cent of the respondents with MRP systems reported that they had a 'Class A' system.

JIT, in contrast, takes the approach of making the problem smaller (sharpening the nail) and using a very small hammer.

Pull systems

Pull systems do what their name suggests: they pull work through the factory to meet customer demands. There is a wide variation in how the concept of a pull system is applied, depending on the characteristics of

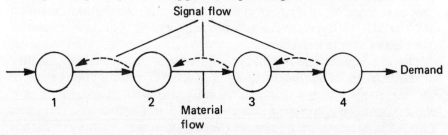

Figure 7.1 *Example of the operation of a pull system*

94 Putting the Just-in-Time Philosophy into Practice

the manufacturing system – for example, whether we are making to stock or to order. Overall they operate as shown in Fig. 7.1 (and Fig. 3.6).

In the example shown in Fig. 7.1, the items pass through the flow line from operation 1 to 2 to 3 and then to the final operation, 4. When there is a demand for an item, it is produced by operation 4 and then taken away. This demand can come from a subsequent process or from a customer. When operation 4 runs short of components as a result of finished products being removed, a signal is sent to the preceding operation (operation 3). Operation 3 then produces components for operation 4. When the component supply for operation 3 is running low it sends a signal to operation 2. The process is repeated all the way through the manufacturing system.

Some points about pull systems ought to be noted:

- machines/operations do not produce any item unless required to by a subsequent machine/operation
- control information flows backwards through the manufacturing system whilst the material flows in the opposite direction
- pull systems help identify problems

The first point (that machines/operations do not produce any item unless it is required by a later machine/operation) is considered to be especially important. There is little point in operations producing items unless they are needed; the items would lie around as work-in-progress on the factory floor. In contrast, the traditional approach to manufacturing control has been to run each machine/operation at full speed even if the subsequent machines/operations cannot handle the output. The result has often been large amounts of work remaining at subsequent machines/operations, increasing manufacturing lead times and work-in-progress levels as well as exacerbating clutter and confusion on the factory floor.

The second point (that control information flows back through the system whilst material flows forward) becomes a factor to be considered when long lead times (usually from suppliers) mean that the system is slow in responding to changes in demand. In this case the addition of an MRP or MRP II-type system may prove beneficial. This is discussed in more detail later in this chapter.

The third point about pull-type systems is that they help identify problems. One of the major features of the JIT philosophy is that we should set up systems that help identify problems. Let us look at a simple example. Supposing that operation 3 in the flow line shown in Fig. 7.1 is a temporary bottleneck. What happens? Operation 3 cannot process as much as operations 1 and 2, so they remain idle for much of the time. This is immediately visible to the supervisor of the flow line and remedial action on operation 3 can be taken. Bottlenecks are frequently dynamic and transitory in nature. The pull systems help to identify and remedy them quickly.

Pull-type systems can enable us to reduce lead times and work in progress levels. There is, however, a wide variation in the manner in which they are implemented. As an example a cloth-dyeing company in southeastern USA has implemented a pull-type system. Batches are taken from machine to machine on carts as shown in Fig. 7.2. In front of each operation a queue of full carts is kept.

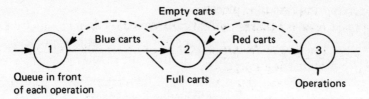

Figure 7.2 *Example of a pull system*

When one operation uses up the contents of a cart the empty carts are sent to the previous operation. This previous operation uses the contents of its *next* input cart (notice the use of first-in-first-out), processes this and puts the completed batch on the empty carts which can then proceed back to the subsequent operation. Carts travel in a continuous loop as shown in Fig. 7.2.; to avoid confusion all the carts in a particular loop are painted in the same colour (red, blue, etc.). This pull system thus operates in a way which maintains a certain buffer queue in front of each machine.

One advantage of pull-type systems that has been indicated is that the work-in-progress levels and manufacturing lead times can be gradually reduced. In the above example this can be done by gradually reducing the number of carts in a particular loop until a problem occurs (equivalent to hitting a rock in the river of inventory – Chapter 3) which is solved, thereby further reducing the number of carts. This is the sort of incremental reduction that has resulted in the Toyota Corolla line having a stockturn of over 80.

Kanban systems

Kanban is Japanese for card. In the Kanban system used in Toyota a card instead of an empty cart is used to indicate to the previous machine/operation that work is needed. For feedback signals other devices can be used such as computer networks, buzzers, or tokens. All

96 Putting the Just-in-Time Philosophy into Practice

that matters is that a machine/operation should receive a signal when work is needed by the succeeding machine/operation. Standardization of containers will enable a standard quantity of work to be shipped from operation to operation. An example is the standardized rack for motorcycle gas tanks at Kawasaki USA.

The Toyota Kanban system differs from the generic pull system described earlier by using two types of signal (equating to two types of card or Kanban). This dual card system uses both withdrawal or conveyance Kanbans and production Kanbans. The withdrawal or conveyance Kanban is used when parts are to be moved between the output and input buffer stocks whilst the production Kanban is used when production is to take place.

The main reason for using a dual card system is that Toyota uses both an input and output buffer store, especially where processes are physically separated, for example, in different plants. This is shown in Fig. 7.3.

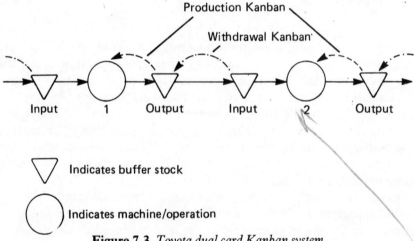

Figure 7.3 *Toyota dual card Kanban system*

As machine operation 2 uses up its input buffer stock a withdrawal or coneveyance Kanban is sent to the preceding output buffer stock and parts are taken from there and delivered to machine/operation 2 input buffer stock. As machine/operation 1 output buffer stock is reduced a production Kanban is sent to machine operation 1 which then begins manufacturing, using parts in the machine/operation 1 input buffer stock. The process is repeated backwards along the production line.

The main advantage claimed for the dual card system is the extra control it gives. However, the extra complexity of the system as compared with the single card approach has led to relatively few companies using it; although Toyota has forced most of its suppliers to adopt a dual card system.

In western industries the single card system is used much more frequently. Another variation is a system based on the use of cards (or other signals) that replenish queues in front of machines (*process* Kanbans) and those that replenish products (*product* Kanbans).

Whichever pull/Kanban-type system is used they can provide a very simple and clear mechanism for shop floor control, 'clear' in that they are readily understood and the factory floor worker is furnished with a simple and visible feedback signal (empty carts or cards). When the workforce at Harley Davidson were told what their new material handling system was going to be, many were amused. Overhead conveyors were ripped out and instead of automated guided vehicles and a complex sophisticated computer control the new system consisted of carts being pushed by hand. However, a result of implementing a pull-type system was that work-in-progress and lead times were reduced by 75 per cent.

Linking MRP with pull/Kanban systems

When properly deigned and implemented pull/Kanban systems can provide an excellent mechanism for shop floor control. Results of actual implementations show that lead times can be reduced and quality levels increased.

However, pull/Kanban systems operate by information flowing backwards and, as indicated, this can be a disadvantage where long lead times are encountered. In these circumstances information takes a long time to be fed back.

The example in Fig. 7.4 shows the lead times for operations 1, 3 and 4 are 1 day each, whereas for operation 2 it is 17 days. This would often be the case where operation 2 was completed outside the JIT shop. Under these circumstances it could take information about demand more than 18 days to reach operation 1 from operation 4. Clearly this could cause many problems if demand changes dramatically.

In such cases efforts should be made to reduce the long lead times, but where this is difficult or impossible (as with specialized raw materials, for example) an MRP system can be used to speed up feedback of information to the early operation. This is shown in Fig. 7.5. The MRP system inputs the future demand and evaluates the requirements for materials and components. The output from the MRP system can be fed directly into the early operations (operation 1 in Fig. 7.5) so that these early operations receive early notification of any changes in demand.

98 Putting the Just-in-Time Philosophy into Practice

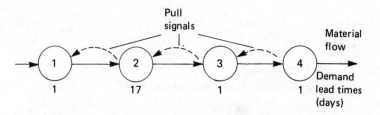

Figure 7.4 *Example of long lead time operation*

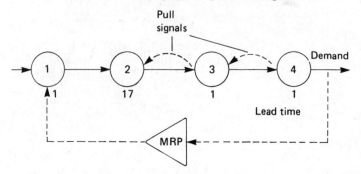

Figure 7.5 *Example of the use of MRP systems to counteract lead times*

In this manner the information for operation 1 arrives, not from operation 2 but from the MRP system. We therefore have a combined push/pull system.

One other aspect for which MRP systems are suitable is coordination. When we have a number of shops (some or all of which may be JIT oriented) the MRP system can integrate the activities of the shops (this process is shown diagrammatically in Fig. 7.6), ensuring that sufficient raw materials/components are delivered.

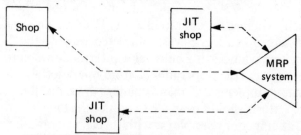

Figure 7.6 *Use of MRP system to coordinate JIT shops*

In both instances (countering long lead times and coordinating shops), the MRP system is not actually being used for detailed control of shop activities. Instead it is used selectively for global coordination. It is this aspect of global coordination to which MRP is more suited and implementations that have used MRP for such a role have usually found that it works well.

Shop floor control and quality at source

The pull/Kanban-type systems provide an ideal mechanism for a systematic attack on work-in-progress and manufacturing lead times. Quality at source programmes can also achieve dramatic improvements. A full description of this aspect is beyond the scope of this book but a useful starting reference is Nagger (1984).

Much of the success of JIT rests on providing the right environment at shop floor level, and for this education is vital (Chapter 5). The education programme should cover all shop floor personnel involved so that they fully understand both the background to JIT and its impact on their own job functions. As indicated in Chapter 5, the most successful JIT implementations are those that concentrate a large portion of the investment in JIT on education. The first stage in installing a JIT shop floor control system should therefore be the creation of the right environment. This is not achieved just by saying 'Everything about the implementation of JIT is straightforward'; indeed this is likely to be counterproductive since when problems are met confidence in JIT will be reduced, and after two or three such problems, confidence will become non-existent. Instead the message should be: 'JIT is great, but it requires a lot of work and it requires us to overcome some problems.' If this message is put across effectively, shop floor personnel are then in an extremely good position to implement a successful JIT system.

What are the changes that JIT brings at shop floor level? They include the following:

- a change in the shop floor culture
- the increased use of suggestion schemes
- the use of productivity circles to improve the operation
- a reduction in the number of command hierarchy levels

We are seeking to introduce a new shop floor culture with a more cooperative and productive environment. This revolves around one important aspect of JIT, which is *respect for all workers*. One obvious reason for this is that it improves the company's image, but an even more important reason is profitability. Respect for all workers makes good business sense. When we are looking for ways to increase productivity and quality, who best knows about a particular process? It is not the middle manager who may spend most of his/her time in meetings, or even the supervisor, who is making sure that materials and tools are available; instead it is the machine operator. He is the one who knows all the idiosyncracies of his machine intimately. If there are improvements to be made, it is the machine operator who will be the most likely to know how to achieve them. Changing the shop floor culture therefore means altering the environment so that the machine operators are fully involved in the continual improvements associated with JIT and so that their expertise is fully exploited. This involves their involvement in quality at source programme using statistical quality control.

This involvement can, in part, be achieved by the increased use of suggestion schemes. In the better Japanese companies, suggestions from shop floor workers are eagerly sought and acted upon (over 80 per cent are implemented). In the West, suggestion schemes have been infrequently used by shop floor workers and those suggestions that have been received are implemented only rarely. But if we are serious about using the expertise of shop floor workers to improve productivity the use of suggestion schemes is vital.

One way in which management can ensure that suggestions will be made and acted upon is by setting up *productivity circles*. These circles are more than quality circles as they deal not only with quality but also with such areas as scrap prevention, reducing set-up time, increasing machine output, material handling changes, etc. (It should, however, be added that many of the better quality circle programmes do already include these aspects.) These productivity circles meet at regular intervals (once a week) and discuss problems and productivity improvement. This is a very good way of encouraging and implementing suggestions. In one company where productivity circles were introduced, each shop floor worker was responsible for one machine. Thus when changeovers to another product occurred, the set-up time was long because the single worker had to do a considerable amount of work. The productivity circle suggested grouping four machines and four employees together and staggering changeovers on the four machines so they did not occur together. The result was that when a changeover did occur there were now four workers available, and this reduced the changeover time by 65 per cent. Furthermore, when they lost one worker through natural wastage they voted that he should not be replaced. This worked well and was repeated over the whole factory. The result was a 25 per cent increase in productivity.

Along with the respect for all workers should go a gradual reduction in the number of command levels in the company. In many companies the number of levels has increased, resulting in more alienation from the factory floor. The president of the $3.5 billion Saturn auto plant, Richard LeFauvre, says, 'In American industries we tend to take executives and put them on pedestals. That keeps executives from knowing what's going on and keeps workers from identifying with them.' In the new Saturn plant executives will be encouraged to eat alongside shop floor workers at a common employee cafeteria as a way to boost morale and to breakdown artificial barriers between workers and management. 'When you eat in cafeterias with everyone else, people will talk to you,' LeFauvre says.

This illustrates some of the changes in attitudes to shop floor workers that JIT has started to bring. Japanese factories operate with fewer command levels than western ones. At one company in the USA which was working towards JIT there were seven levels between the plant manager and the shop floor worker. In the equivalent Japanese factory there would probably be only three. Reducing the number of levels in the hierarchy reduces the sense of alienation from the top management that is commonly felt by the average shop floor worker and, in the long run, will gradually reduce the number of indirect employees through natural wastage. 'White collar' productivity often shows a large increase in JIT implementations.

This whole area of shop floor culture is the one that can, if it is got right, lead to continued improvements in productivity or, if mishandled, lead to continued disappointments. Creating the right culture for JIT is therefore one of the major milestones in a successful JIT implementation.

Conclusion

Step 4 marks the final step of in-company activities associated with JIT implementation. The way in which the JIT factory is controlled will largely determine the value of the benefits that JIT will bring. This chapter has described how it should be approached and the control mechanisms that are most commonly found in a JIT environment. The basis of JIT control, pull/Kanban system, is simple and its operation is clearly visible to factory floor workers. It is a system that, by lowering the work-in-progress levels, gradually identifies the problems. When the problems have been identified they can be dealt with and waste will be eliminated. In this manner pull/Kanban systems address all four of the main features of the JIT philosophy (Chapter 3).

Pull/Kanban type systems can be divided into the single card and dual card systems. The original Toyota implementation used a dual card system which gives greater control but is more complex than the single card system. It is for this reason that the majority of pull system implementations use the single card system, although other mechanisms besides cards are often used.

Besides establishing a control methodology (such as pull/Kanban type systems) it is also necessary to create the right environment on the factory floor to install quality at source or statistical quality control programmes. This can be done by:

- changing the shop floor culture
- increasing the use of suggestion schemes
- using productivity circles
- reducing the number of command levels in the hierarchy.

Companies with the most successful JIT implementations have completed all four tasks. In so doing they created an environment that is both cooperative and constantly seeking improvements. If their example is followed the benefits from JIT will continue to increase year after year.

8
STEP 5: VENDOR/ CUSTOMER LINKS

Introduction

Step 5, vendor/customer links, marks the final step in the implementation of JIT. Having completed steps 1–4, a company wanting to gain maximum benefit from JIT must now take a wider view. So far the focus has been on in-house changes intended to improve the manufacturing process or to improve manufacturing control. But a manufacturer can only gain so much benefit from in-house changes and in order to give scope for further improvements changes in vendor/customer links may have to be suggested. These changes may well emerge as critical issues in the final stages of the implementation.

Both the vendor and customer links are important in the JIT implementation in that they broaden the scope of a cost reduction and the drive for quality improvement. For example, if we take steps to raise the quality of our vendor's components this both reduces the measures that have to be taken when a large, poor quality batch is received and ensures that the improvements in the quality of components produced in-house will be matched by comparable improvements in components, from outside suppliers, resulting in a higher quality final product.

Savings can be large. Recent research suggests that in western industries material costs account for 51 per cent of total costs, whereas labour costs account for only 15 per cent. Trends for labour costs, as a percentage of total costs, are being reduced (in many industries labour costs are below 10 per cent of total costs), whilst materials costs tend to be increasing. Technologies such as automation and robotics have reduced labour costs and many companies are making large investments that will reduce them even further.

By contrast, companies are only just beginning to examine areas that can significantly reduce material costs. Purchasing departments have frequently been content to take a short-term view, responding to changes in demand, scrap or obsolesence with urgent requests to vendors for expedited orders.

Customers are important because, from the financial point of view, they supply the money and, from the manufacturing management point of view, they drive the whole manufacturing process. Obviously, with no customer demand there would be no manufacturing.

If customers are brought into the JIT implementation both the customer and the manufacturing company gain. For example, if the customer can give a schedule of firm orders for a particular time ahead (often 6–8 weeks), then the manufacturer with the short manufacturing lead times often associated with JIT can work to this schedule in the knowledge that no changes will be made, thereby reducing costs. Part of this cost saving can be passed to the customer. Additional benefits may occur as there is more time available to concentrate on quality.

Harley Davidson, the motorcycle manufacturers, included vendor/customer links in their JIT implementation. Instead of the traditional adversarial relationship with vendors and customers they established a more cooperative one; the change in attitude was demonstrated by their description of vendors/customers as being 'partners in profits'. This stressed the advantages to the vendors and customers of being involved in the JIT implementation. JIT will achieve benefits if implemented within the company, but will however achieve far greater benefits if the vendors and customers are included in the implementation.

This chapter describes some of the essential features to be considered when including vendors/customers in a JIT implementation. These include:

- links with vendors
- multi-sourcing versus single sourcing
- short- versus long-term agreements
- local versus distant suppliers
- how to implement links with vendors
- links with customers

Step 5, like the other four key steps, requires a good deal of pre-planning, time and effort.

Links with vendors

Purchasing has often been a neglected area of management, but it is in purchasing that we can make substantial cost savings; on average, for every dollar spent on labour over three dollars are spent in purchasing. Our scope for cost reduction is therefore greater in purchasing than in labour costs (although we should not ignore that area either!).

How does the JIT approach effect the relationship with vendors? First, we should remember one of the four aspects of JIT philosophy: *eliminate waste*.

With vendor links one way of reducing waste, in the form of surplus inventory, is to reduce the order quantities. Why? Suppose we use 1,000 of a particular component every week and the vendor delivers in batches of 4,000 every four weeks. The *average* time a component is in stock is two weeks. If we reduce the order quantity from 4,000 to 1,000, deliveries are made every week and the average time a component is in stock is reduced to half a week.

Reduction in order quantities is one aspect of JIT that is applied to vendors but some changes are necessary to make it workable:

- minimize bureaucracy
- rim deliveries
- streamline inventory management

In the example above we reduced the order quantity from 4,000 to 1,000. Since the requirement remains at 1,000 per week, there are now four times as many deliveries as previously. This can only make economic sense if we alter some of the vendor mechanisms. First we need to streamline the associated bureaucracy to cut down on the paperwork. If each delivery has the same amount of paperwork it will increase when there is one delivery a week. This can be cut down, for example, by only sending out one order every month but scheduling shipments within that order daily or weekly.

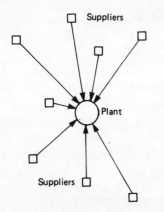

Figure 8.1 *Spoke system of deliveries*

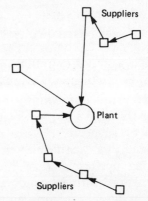

Figure 8.2 *Rim system of deliveries*

More deliveries can also lead to higher freight costs as more journeys have to be made. Figure 8.1 shows deliveries made on a spoke basis, each supplier delivering directly to the plant. To reduce the cost for shipping smaller volumes a rim system can be used (see Fig. 8.2). Vendors take turns to deliver to the plant, calling at other vendors on the way. For large-volume vendors, direct links could be kept where this is justified. This rim-type system requires some organization but does have the potential to lower the cost of shipment.

When the items arrive in the plant, the inventory management should be streamlined so that the items go quickly on to the shop floor where they can be used. This means cutting back on inspection and the receiving goods inventory. These important procedural changes are offset by quality improvements which, for example, eliminate the need for in-goods inspection.

Streamlined bureaucracy and inventory management, together with rim deliveries, are some of the changes needed to facilitate JIT vendor links. Major requirements of JIT vendor links are:

- high quality levels
- reduced order quantities
- reduced and reliable lead times

These help to reduce inventory levels and uncertainty about vendor performance. If we can be certain a vendor will deliver high quality items on time our safety stock can be reduced, together with the need for in-coming inspection, and there will be no disruption to production through poor quality items or through late delivery.

Vendors can be classified into categories depending on their delivery performance, as:

> On-time delivery: A, very good; B, good; C, poor
> Delivered quality: 1, very good; 2, good; 3, poor

These then give an indication of their delivery performance. For example, a C1 supplier is one who consistently delivers high quality components but whose on-time delivery performance is poor. Similarly, an A3 vendor is one who delivers on time but whose product quality is suspect.

Using this classification scheme, many companies initially take the existing performance of the suppliers to categorize them. During negotiations with the vendors, it is made clear that they will be expected to be in the A1 category but that as time goes by the specification of the A1 category will increase in severity. It is important to note that eventually *all* vendors are expected to move into the A1 category, and to stay there. This method is a useful way of aggregating complex data into an understandable form.

The costs associated with a purchase order can be divided into six types:

1. negotiation costs.
2. changing a planned order to a firm order and associated paperwork.
3. expediting costs.
4. receiving-count costs.
5. receiving-inspection costs.
6. premium freight costs, ordinary freight costs.

With JIT the total cost is not viewed as fixed, but each element is dealt with in turn as part of the campaign to eliminate waste.

When a vendor meets category A1 requirements (provided these are strict enough), we can then substantially cut back cost elements 2–6. In addition, the negotiation cost can probably be reduced. But when a vendor is rated A3, then only the costs 2, 3 and 6 are substantially reduced.

If our aim is to have *all* vendors in the A1 category then we must be prepared to enter into long-term agreements in a spirit of cooperation. This cooperation should include a careful move towards single sourcing, long-term agreements and local supplies. In choosing a vendor we are probably moving away from a selection process based entirely on primary cost towards one based on total cost, which includes the cost of factors such as late deliveries and a reject batch. When all the costs are taken into account it may be that a vendor with a higher purchase cost has an overall lower total cost as disruption and cost caused by late or poor quality deliveries can be substantial.

Small companies who purchase from large companies may encounter severe problems in attempting to improve quality and delivery performance. Options in this case include moving to other suppliers or persuading the larger companies to change.

Multi-sourcing *versus* single-sourcing

Conventionally, most large manufacturers have relied on multi-sourcing components. This involves several vendors producing the same part. The advantages include greater security of supply (any disruption with a single vendor does not stop the supply) and cost reduction (through greater bargaining power). However, those who argue in favour of multi-sourcing can overlook three critical points.

First, they may ignore the economies of scale. If a vendor can supply greater quantity a reduction in cost follows as many of the fixed costs remain the same. Second, each vendor deals with smaller volumes than with single-sourcing, and this volume may not necessarily be sufficient to justify future investment in improved processes. Third, there are more managerial problems associated with dealing with several vendors.

The JIT approach stresses the need to move towards using a single source. Indeed, it continually emphasises the need to have a single vendor supplying several parts in a 'family' thereby increasing the volume per vendor and reducing the number of vendors. This will encourage the vendor to make the capital investment required to improve his manufacturing processes. Frequently, large companies implementing JIT will send a task team to vendors (especially small vendors) to study their manufacturing processes and to recommend changes.

Does reliance on a single vendor raise the potential problem of supply disruption? If the single source vendor is disrupted does this mean that we are more likely to have our own supply disrupted? The answer to both question is a tentative 'no', as long as the groundwork has been done well. This includes building up a good relationship with vendors, giving them technical advice when necessary and ensuring that the vendors chosen are financially secure and well managed. If they are not, another vendor should be selected. It is extremely important to move slowly and carefully towards single source supply. Obviously relying on an unsuitable vendor can be disastrous to company operations.

Short- *versus* long-term agreements

Traditionally, long-term agreements have been viewed with a good deal of suspicion by purchasing departments. It means tying the company to a particular vendor over a long period of time with little opportunity to re-negotiate or seek alternative vendors. Purchasing has often welcomed short-term agreements because they allow more flexibility and more competition on price. At the end of a short-term agreement, re-negotiations can take place with several vendors, with those quoting the lowest price getting the contract.

From the vendor's point of view long-term agreements are preferable as they involve less risk. Short-term agreements can involve greater costs because they provide the vendor with no incentive to invest in improved processes and so reduce costs.

JIT encourages long-term agreements with a few carefully selected vendors, for the following reasons.

- more reliable deliveries
- greater opportunities for investment
- better quality products
- lower cost

A vendor with a long-term agreement is thought to be more likely to keep to delivery promises, frequently at the expense of short-term agreements with other companies. The company now becomes a major customer (especially if the company is purchasing a product family from the vendor) and their demands will be met first.

The long-term agreement is also thought to give the vendor a greater sense of security. It is therefore appropriate for the vendor to make some investment to help the production of the product family, investment in machinery, control systems or in training personnel.

Part of the long-term agreement will specify the due date and the quality levels. Before JIT, when contracts were often awarded solely on the basis of price, there was perhaps little incentive for the vendors to improve the quality of their products.

Some gave up even trying to deliver items of an acceptable level of quality. The long-term agreement specifies required quality levels (usually at an increasingly high level), with vendors left in no doubt as to the importance of maintaining this high quality.

These investments together with the higher production volumes for each vendor lead to cost reduction, part of which benefits the vendor and part of which is passed on to the purchaser.

However, it is not advisable when implementing JIT to immediately rush out and make long term agreements. One of the reasons why the vendor/customer link is Step 5 and not Step 3 or 4 is that it takes time to identify and establish a relationship with suitable vendors. One cannot change overnight from the traditional, sometimes antagonistic, relationship with vendors to the ideal JIT environment or trust and cooperation. Time is needed and the groundwork for vendor/customer links can be laid whilst steps 2, 3 and 4 are in progress. The aim should be a *gradual* move to long-term agreements. Any purchasing manager who overnight makes long-term agreements with single-source, higher volume vendors is putting the company's prospects and his own career in jeopardy.

Such agreements should only be entered into after considerable thought and analysis. Companies that have successfully implemented the JIT approach with their vendors have gradually extended the agreement period and moved slowly but steadily over to single-source, high volume vendors.

Local *versus* distant suppliers

A large micro-computer manufacturer recently found an entire month's shipment of nearly half a million disk drives which had arrived from Japan in a single shipment to be faulty. They had to be sent back to Japan (five weeks delay), disassembled, rectified and returned to the USA (another five-week delay).

Such occurrences are not unusual in long-distance purchases. The long transport time adds significantly to the problem. For the micro-computer manufacturer the cost and disruption of the faulty disk drives could have been minimized if the vendor:

- delivered in small lots
- was local

If, instead of delivering nearly half a million (a month's supply) in one lot, the vendor had delivered in lots of say one, two, or three day's supply, the problem could have been identified when a much smaller number of disk drives had been made. The problem would then have been much less severe. If, in addition, the vendor had been local the five-week transportation time would have been avoided. The fact that transportation costs are rising at a faster rate than many other costs is increasingly powerful economic argument in favour of local vendors. In addition, the long delivery times associated with distant vendors reduce flexibility. Every day added to the manufacturing lead time for transportation extends the planning horizon. For example, if the manufacturing lead time for the disk drives is one week and the delivery lead time is five weeks, the disk drive vendor must be told six weeks in advance how many disk drives are required. Reduce the delivery lead time to a few hours and the vendor need only be notified one week in advance. This can greatly reduce problems and uncertainty.

Local vendors therefore reduce waste through inventory associated with the delivery time and lessen the risk of a large defective delivery. In addition, risk and uncertainty associated with long lead times are reduced, thereby making the system more flexible at a lower cost.

How to implement links with vendors

The implementation and maintenance of links with vendors are very important to the success of a JIT system. As indicated pre-planning for improved vendor links should be carried out in parallel with part of step 2 and steps 3 and 4. During this pre-planning phase, suitable vendors are identified and longer term agreements are made with some selected vendors. By selecting only a few, if any, single-source vendors at this stage options are kept open.

The overall aim of the pre-planning and the initial stages of step 5 should be a gradual identification of suitable vendors and a gradual implementation of a single-source, higher volume purchasing. During step 5 a large amount of time is spent in direct contact with vendors. I have found that an initial one-day meeting/seminar with the vendors works well. Plans for the introduction of JIT and the implications for future dealings with vendors are described. Frequently, the vendors have had other customers who are implementing JIT or else they themselves have briefly looked at JIT and are familiar with it. The vendors should be left in no doubt that what JIT requires of vendors is:

- a frequent, high quality supply and
- high due date achievement which can lead to
- an A1 classification.

It should be made clear that any vendor who attains A1 classification can be sure of a long-term agreement which should generate enough profit for the vendor to invest in new process equipment and in training personnel. The vendor will then be better placed to achieve the high quality, small lots, and high due date achievement required. The requirement is for all the vendors selected to be in the A1 category and for specifications within this category to improve gradually.

A task team will frequently be sent from larger companies to the vendor plants to advise vendors on how small lots with high quality and short, reliable lead times can be produced. This is especially useful for smaller vendors who may not have the necessary expertise.

The agreements themselves are long-term, with six months to one year being a definite commitment; any period beyond one year is an informal agreement. The vendors can, if certain conditions are met (such as keeping to the A1 classification), be relatively certain that their contracts will be renewed.

Overall, step 5 consists of eradicating the barriers between the company and its vendors and forming definite links leading to long-term agreements. The result in the companies that have achieved this seems to be an increased quality and an increased due date achievement of goods supply as well as lower costs for both the purchaser and the vendors.

Links with customers

Forging links with major customers is the last part of the chain of JIT which passes through vendors, the company and on to the customer. It is important to include customers in a JIT implementation as their input can ease planning problems. If, for example, a major customer gives a firm schedule of his requirements six weeks in advance and the company lead time is five weeks then the schedule can be met relatively easily. If the customer only gives a firm schedule one week in advance then the company encounters more problems.

The major function in forging links with customers is one of education. The customer must begin to realize that if a firm schedule is given some weeks in advance and not changed then they can be relatively certain that the schedule will be met. This reduces costs and disruption both for the company and for the customer.

Again a one-day meeting/seminar with the major customers, explaining JIT and why early feedback of customer requirements is needed, can help. From the customer's point of view a supplier who implements JIT may well have reduced lead times (giving a good response to changes in demand) and improved quality. One JIT implementation which I have been closely involved with began one year ago and has not had a late delivery in six months. It is this kind of benefit that customers appreciate, but they should realise that they have to supply a firm schedule. Again, a long-term agreement may help the company. Overall the aim in forging links with customers is to improve the response of the JIT system to changes in market requirements. This can in turn reduce costs to the customer. I have frequently (though not always!) found that when the basis of JIT is presented to customers in a fair and objective manner they will be enthusiastic to make the necessary changes, which are often only slight.

Conclusion

Improvements in vendor/customer links only make sense when a large portion of the in-company JIT changes have been made. But step 5 should begin in parallel with part of step 2 and with steps 3 and 4, for time is needed to discuss the requirements of JIT with vendors and customers and the necessary changes take time to implement.

With JIT the company considers using fewer suppliers, these vendors generally being single-source and each producing higher volumes. The move to a single line of supply should happen gradually to ensure that the right vendors have been chosen, for any purchasing manager who changes to single-source, higher volume vendors overnight is perhaps putting the future of the company and with his/her own career in jeopardy.

I have found that those JIT implementations that forge good links with vendors and customers gain a significant amount. The net result is a higher quality, lower cost supply that is delivered on time with greater security for both vendor and customer.

9
JIT IMPLEMENTATION – THE PROVEN PATH

Introduction

The way in which we implement steps 1 to 5 will be reflected in the results that we obtain from the JIT implementation. The questions usually asked after examining steps 1 to 5 are 'O.K., these are the procedures, but how should we implement them?' 'How long should the implementation take?' 'What should be done to help implementation?'.

Fortunately, there are a growing number of relatively successful JIT implementations, some of which I have been involved with. From these implementations the five steps and a proven path to implement these steps can be derived. A proven path consists of an implementation sequence and timings for each step. Drawing on this experience, any company wishing to implement JIT can follow each step in the correct sequence with the timings given.

Many of the detailed points concerned with the implementation of each step have been given in the chapters describing them. This chapter gives an overview of the timings and sequence in which the steps should be implemented. Probably no one company will ever follow this model to the letter, but it will provide a blueprint against which a particular implementation plan can be compared.

Each of the five steps is crucial to the eventual success of the JIT system. One can compare implementing JIT to building a home:

Step 1 making a commitment; choosing and buying the lot
Step 2 putting in the foundations
Step 3 building the framework
Step 4 putting the roof on
Step 5 finishing the interior

You will still end up with a home even if every step is not done properly. However, if you want a good quality house, each step has to be done properly and in a certain sequence (there is no point in building the framework until the foundations have been put in). Timings should be realistic (too short, and there is a risk of poor workmanship, too long and some parts may deteriorate with age and weather). In the same manner a good JIT implementation requires that all of the five steps described in this book be done well. Time allotted for each step should be sufficient to ensure that it is accomplished properly but not so long a time that enthusiasm starts to deteriorate.

JIT is important to the future of the company. The gradual spread of the JIT philosophy from a small group of pathfinders to encompass the whole company has been described in this book. There is no doubt that JIT, when properly implemented, can be successful. Some results from JIT are:

- One Hewlett-Packard plant (manufacturing small disk drives) reduced inventory from 22 days to just over 1 day in 18 months.
- Harley Davidson reduced inventory by 75 per cent; warranty costs have been cut by 30 per cent.
- In one General Electric plant inventory was reduced by 80 per cent and quality improved by 50 per cent.
- At Xalloy Inc sales more than doubled because the right products were available at the right time.

The costs associated with each of the implementations are low relative to the returns.

How long should the implementation take?

The timescale for any project can be reduced by increasing the resources allocated to it. This applies to some extent to JIT. If we shorten the time scale for implementing JIT we obtain the benefits earlier. There is therefore some pressure to implement JIT in as short a time as possible. Since the costs associated with are JIT often relatively low there is normally no problem obtaining extra resources if it is obvious that they will reduce the implementation time scale.

However, JIT is a philosophy that requires a fundamental change in behaviour, so even if all the necessary resources are made available a company could not implement JIT in a short time. The changes required are so fundamental that it takes time for the implications to register and for the philosophy to be absorbed into the day-to-day running of the company.

Overall, companies that have successfully implemented JIT have found a time scale of one-year for implementation to be ideal. Why one year? First, as the financial resources required for JIT are usually not high there is generally little problem in obtaining them. Second, a shorter time will not enable the necessary changes in attitudes and philosophies to take place. Third, if longer than a year, there is a risk that enthusiasm for JIT within the company will wane since results will take longer to arrive. Furthermore, with a time scale longer than one year, key personnel are more likely to move on or be transferred from their posts. Their replacements will need to be educated to implement the JIT process further which takes time and can cause a slight setback. There is still a risk of personnel changes in a one-year implementation plan but this is obviously likely to be less.

By an implementation time scale of one year I mean from the *end* of step 1 onwards; that is, after the decision to proceed has been made and the project team appointed. I have excluded step 1 because I have found it varies the most in time scale. From the pathfinders' first queries about JIT until the project team is appointed only takes a few weeks in some companies, in others it can take several months. There is no fixed time scale for step 1; the important thing is to get started properly.

Step 1 should not be rushed in order to get JIT implemented as quickly as possible. Not completing each phase of step 1 can result in problems later in the implementation, which can affect the whole implementation's progress. In case study B, described later in this chapter, the pathfinders did not have full commitment of top management. This hampered progress all through the implementation of JIT which would have been more successful with this commitment.

Having completed step 1, the company arrives at the start of the implementation. The start date we can call D-Day since many companies consider it helpful to view implementing JIT as a military campaign with the commitment that this involves. Our goal in this case is to have JIT implemented one year after D-Day.

I will stress again that the implementation of JIT is just the first stage. What JIT should mean is a process of continual improvement maintained long after implementation has been completed. The implementation puts in place an infrastructure which should allow improvements to occur for years to come – but this will only happen if we ensure that the company does not revert to previous operating procedures. An on-going education programme is therefore vital together with a total acceptance of JIT by top management.

The implementation sequence – the proven path

The sequence that has been found to work best in JIT implementation (the proven path) is shown in Fig. 9.1 together with its time scale.

This is a one-year implementation time scale, from D-Day, with many of the steps occurring in parallel. We can schedule the steps in parallel because most of the steps require different personnel. For example, step 5 – vendor/customer links, mainly involves purchasing and sales personnel, whereas step 3 – process improvements, involves industrial/manufacturing engineering and shop floor personnel.

The steps can be either in full progress (in foreground mode) or in a background mode, and for each step the periods in foreground and in background modes are shown in Fig. 9.1. Step 1 precedes D-Day so that on D-Day we are ready to undertake the subsequent steps.

Step 2 – education is the foundation step. Without it the implementation would be very poor with a high likelihood of failure. Step 2 therefore begins on D-Day – as soon as education is underway the other steps can follow. The education programme will take around seven months to cover all relevant personnel. After the seven-month programme of initial education has been completed, we now put step 2 into background mode where education is on-going but is more a refresher-type format with more time allocated to a discussion of problems and their solution rather than to a conventional seminar format.

Step 3: process improvements involves improving machines and processes so that the correct quantity of good-quality produce is made. Such improvements involve reducing set-up time and improving machine reliability. This step can begin in background mode in month 1, move quickly to foreground mode in month 2 and continues in foreground mode until month 9. Process improvements can still continue after month 9 but in a background mode.

Step 4: control improvements includes the use of pull/Kanban type systems and changes in the decision-making hierarchy. Some of these changes are fundamental and will require virtually until the end of the 12-month period to implement. The move to a more pull/Kanban type system cannot begin until some of the process changes (step 3) are in place. Consequently step 4, although largely in parallel with step 3, runs a little behind it.

Once many of the in-house changes (steps 3 and 4) are in place, step 5, vendor/customer links can begin in earnest. Much of the pre-planning and many meetings with vendors/customers are best started earlier (in month 2) but the main activity (foreground mode) is

118 Putting the Just-in-Time Philosophy into Practice

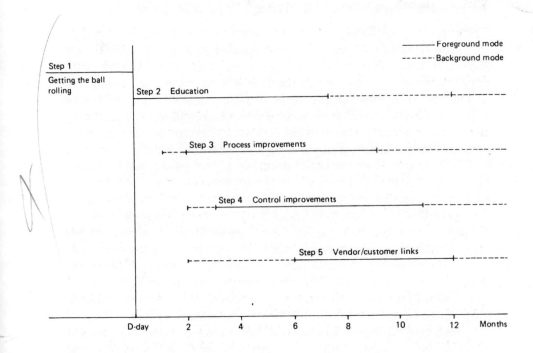

Figure 9.1 *The proven path to successful JIT*

best begun in month 6. The reason for this is that there is very little point in having, for example, just-in-time deliveries from vendors if these lie around in the factory for a long time. Just-in-time deliveries only make sense when the factory is actually operating a just-in-time system. Building relationships with vendors/customers is a very time-consuming business and cannot be achieved overnight. It is to be expected that substantial improvements in vendor/customer links will continue long after the 12-month implementation period is over.

Overall, the proven path (Fig. 9.1) provides a viable and effective JIT implementation plan. Each of steps 3 to 5 requires some fundamental changes in behaviour and hence they are all dependent on step 2, education, which has been described, very accurately, as the make-or-break issue.

The operation of the proven path is now illustrated by case studies. These involved companies who eventually achieved successful JIT implementations, mainly by following the proven path; where it was not followed, problems did result.

Case study A

Company A manufactures consumer durable equipment in USA. The products are mainly for US consumption although there is a significant export market. Until the late 1970s the company faced little competition for its products and enjoyed healthy profits.

The first competitive products from Japan appeared in 1979 and by 1981 ten competitor products from Japan had appeared on the market. The company's penetration of the market dropped from nearly 80 per cent to little more than 30 per cent and it found itself in deep trouble. It had fixed capital assets and large fixed costs yet its sales were dropping sharply. The company was in grave danger of filing for bankruptcy.

A senior management study trip to Japan initiated enthusiasm for JIT. However, what impressed the company's managers even more was visiting a Japanese assembly plant in the southern USA. The plant had taken raw rural labour and educated and trained them in using JIT. The plant was achieving higher levels of efficiency than the equivalent plant of company A. Yet company A had a much more highly skilled workforce. Seeing the high efficiency levels of the Japanese assembly plant in the USA produced even greater enthusiasm for JIT.

Company A then proceeded to complete step 1 of JIT quickly and professionally. A commitment was made, a decision was made to go ahead and the task team and pilot plant identified. The pilot plant chosen was judged to be the most difficult of the company's plants to control. 'If JIT can work in that plant it will work anywhere!' was a typical comment.

A detailed plan for the JIT implementation was produced and the education programme began. By the time step 3 was started enthusiasm for JIT had mounted. The plant layout was changed so as to have small flow line cells instead of the cumbersome layout that was in place. A set-up time reduction programme was also begun. Industrial and manufacturing engineers studied each process of set-up, talked with operators and videotaped the set-up operation. By studying the set-up operation in detail, the engineers were able to significantly reduce set-up time.

All the measures used to reduce set-up time had two things in common. They were simple and cost very little to implement. The result was a 75 per cent set-up time reduction, with one further, unexpected, advantage materializing in that the set-up also became more accurate. Instead of a long adjustment period when the machine was gradually adjusted to a new product, there was now a much quicker adjustment. The result was a 90 per cent reduction in scrap which had

120 Putting the Just-in-Time Philosophy into Practice

previously occurred during the adjustment phase immediately following set-up. The accuracy of the set-up also led to better quality of products. Warranty costs were reduced by 30 per cent, thanks to the action taken on set-ups and other measures such as the introduction of a pull/Kanban type control system.

After step 3 had begun, step 4: control improvements was initiated. Prior to the implementation of JIT the company had a sophisticated MRP system linked to overhead conveyors and a high-rise parts storage. JIT implementation involved eliminating these high technology tools and replacing them with pushcarts.

The pull system used is an adaption of the general pull/Kanban system described in Chapter 7. Company A decided to use a three-pushcart system for replenishment of parts, with the assembly operation of the three-pushcart system shown in Fig. 9.2.

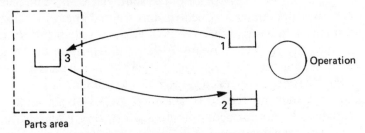

Figure 9.2 *Use of three-pushcart system*

Each cart contains parts for about one hour's production. While the operator takes parts from a cart a second cart is waiting next to him with a complete load of parts and the third cart is being filled in the parts area. When the first cart is empty it is taken to the parts area for filling and replaced with the third cart.

Each cart has a card (equivalent to a Kanban card) which describes the part type and the number of parts per cart. The workers in the parts area examine the card and fill the cart accordingly in a similar manner to the operation of a two-bin system (Chapter 2).

Before the use of the pull/Kanban system, the scheduling of parts and production had been a serious problem ('a nightmare,' according to the plant manager). Items were produced in batches and one of the biggest problems was ensuring the right parts were at the assembly line at the right time. Another problem was the difference in manpower requirements for each product. The more complex products required 140 men to build while the less complex required 90. People were therefore moved on and off the assembly line when different batches moved through.

Step 5: JIT Implementation – The Proven Path   121

A large amount of time was spent discussing the implication of the JIT system with the workforce. Each employee affected by the changes was consulted and many ideas came from the meetings. One of the major recommendations was to change the method of assembly. Under the old method each product was assembled in a batch. For example, a complete batch of X was assembled before batch Y was started (see Fig. 9.3). Since each batch had different manpower requirements, workers were put on or taken off the assembly line as and when required.

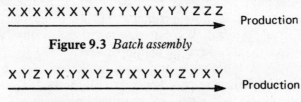

Figure 9.3 *Batch assembly*

Figure 9.4 *Mixed model assembly*

Mixed-model assembly (see Fig. 9.4) involves moving away from a batch mode towards a *pattern* of assembly. Consider, for example, the assembly of three products, X, Y and Z, in the volume ratio 2:3:1. For each product Z assembled, 3 Ys and 2 Xs are assembled. A suitable pattern for this would XYZYXY, repeated again and again. The pattern requires a certain workforce level and since the pattern is repeated the workforce level on the assembly line is kept constant. This overcomes to a large extent the problem of workers being moved on to or off the line.

The products are manufactured to a monthly schedule which remains fixed. This helps vendors plan ahead for capacity. However, the day-to-day supply from the vendors is controlled by Kanban cards, which are posted to the suppliers when production is needed. Therefore, although day-to-day requirements can vary, the suppliers operate within a global monthly fixed schedule.

The MRP system is still used to set the product mix but individual shop orders are now set manually. This illustrates the point that MRP is perhaps best applied in *global*, not *detailed*, control.

The relationships and agreements with vendors are expanding. At the end of the first year 22 of the vendors had been included in the JIT system, 40 by the end of the second year and almost all of the company's vendors had been included by the end of the third year. As stressed in Chapter 8, the links with vendors and customers can be slow to emerge and should not be rushed. Company A have found that by a steady improvement in their links with vendors/customers they can achieve lasting benefits.

The overall improvements from JIT have been impressive:

- inventory reduced by 50 per cent
- set-up time reduced by 75 per cent
- set-up scrap reduced by 90 per cent
- warranty costs reduced by 30 per cent
- high rise parts storage has been eliminated, saving $500,000 annually
- quality improvements

The company, which was near bankruptcy at the start of the JIT implementation, is now profitable again. It has also improved its competitive position: 'We can now stand up to our Japanese competition – face to face,' says the plant manager.

Company A achieved a successful implementation by following the five steps described in this book. It still views JIT, three years after implementation, as a source of continual improvement. It still surprises top management that all of the benefits came about for very little cost. There was no expensive computer hardware or software, no robots or machine tools to purchase (though these can help in certain circumstances). Instead JIT has shown that large benefits can accrue through changes in attitudes and procedures.

Case study B

Company B, in the eastern USA, manufactures metal components that are used in manufacturing equipment. The company had enjoyed a period of tremendous growth. In 1968 it had 65 employees and an annual revenue of $2.5 million. By 1980 the company had grown to 255 employees, annual sales of $11 million and controlled 70 per cent of the world market for its products.

This growth was led by a general manager who had a high school diploma and who had acquired his expertise whilst working. He therefore had a great deal of practical experience but did not feel at ease with the increase in financial expectations and the growth in technology, especially computer technology.

The recession of the early 1980s hit Company B hard. Foreign competition began to affect the market and sales fell to $8.5 million. The company had to make 100 employees redundant and began to operate at a loss.

The plant manager and production controller discovered JIT, and soon afterwards decided to implement a JIT system. The normal pattern of step 1 was followed with one important exception – there was no top management commitment.

Step 5: JIT Implementation – The Proven Path

The general manager felt at home with the more traditional approach and as implementation began he became unhappy with some of the implications. The result was employees did not receive proper instructions and found themselves dealing with a management that became more divided as the implementation progressed.

At the start of the JIT implementation, company B had a number of problems, the most significant of which was excessively long lead times in-house. Shipping 200 products a month, the company often had 2,000 products on the factory floor. Changes in-house therefore became a priority.

Machines were placed in small cell flow lines to cut down the distance travelled by the products; processes were examined in detail and improvements were made. At the start of the JIT implementation, the expeditor was the main production control mechanism. A job would be found on the shop floor and the expeditor pushed through subsequent processes to get it to a customer. The result was general confusion with work-in-progress stacked everywhere.

The JIT system quickly began to show some results; manufacturing lead times were halved. As work-in-progress levels began to drop some workers showed resistance. The educational programme was not as comprehensive as it should have been and some workers did not understand that a drop in work-in-progress was beneficial; they associated high work-in-progress level with the prosperous 1970s. When the work-in-progress levels dropped they assumed that another drop in sales had occurred. Rumours of employee cutbacks also started to circulate, and signals from management were mixed. On one occasion the general manager ordered a six-month supply of raw material in spite of JIT suggesting very small order quantities, and confidence in JIT fell.

Nevertheless, the implementation progressed well with a sharp increase in productivity occurring. Relationships with vendors and customers were strengthened. At the end of the year-long implementation, 90 per cent of the total dollar purchases were with JIT vendors. Company B now supplies its vendors with extensive information concerning its own forecasts of demand, and deliveries to the company usually occur weekly. The results has been that the raw material stocked at Company B has halved while the prices for raw material have also dropped.

Relationships with customers have been pursued aggressively. Reduction in manufacturing lead time has meant that the company is now in a competitive position with regard to overseas competition. Furthermore, many customers are themselves starting to implement

JIT and strengthen their own customer links. Again, the results have been impressive: sales have doubled while the labour force has only increased by 10 per cent.

The board of directors saw the results and future prospects of JIT but also noted the lack of top management commitment. Eventually the general manager was forced to take early retirement and was replaced with a manager enthusiastic about JIT.

Overall, the results of the JIT implementation have been considerable:

- sales have doubled
- manufacturing lead times have halved
- work-in-progress has been reduced by 80 per cent
- raw material prices are lower
- productivity has nearly doubled

The company is now very profitable and is planning further expansion. The problems that it had with the JIT implementation have come from two sources:

1. a lack of top management commitment (step 1).
2. an education programme that was poorly planned (step 2).

The company still managed to implement JIT relatively successfully in spite of its problems but everyone associated with the implementation agrees that it would have been more successful as well as being easier, if these two problems had been dealt with earlier.

Conclusion

This chapter has presented an overall implementation plan for JIT. From the growing number of successful implementations we can distill a proven path consisting of the timings and sequence of implementation of the five steps described in this book. Each of the five steps, together with a carefully thought out implementation plan, is vital. A company that is implementing JIT needs to follow the steps described in this book and to implement the steps using the proven path.

An important point that has been stressed throughout this book is that JIT aims for continual improvements. Improvement should not end once the implementation is complete. JIT implementation is therefore a process of setting up the infrastructure for continued improvements. Prerequisites for this are a proper assimilation of the JIT philosophy and an on-going education programme.

The two case studies amply demonstrate the potential benefits of JIT implementation. Company A followed the five steps and the proven path almost exactly. A highly successful implementation resulted. Company B followed the five steps and the proven path only generally; there were some failings in top management commitment and in the education programme. The resulting implementation therefore had some problems although it, too, was successful. It would have been an easier implementation, and a more successful one, had the five steps and proven path been followed more exactly.

10
SUMMARY AND CONCLUSION

Just in time systems

The challenge western manufacturing industry faces from offshore competition is immense. In several countries whole industries have been eliminated. Manufacturing management must therefore change considerably in order to meet these new challenges and to ensure future viability. Many experts are convinced that manufacturing management approaches will change more over the next 10 years than they have over the last 100 years. Central to this change is the application of Just In Time systems.

As has been stated throughout this book, a successful JIT implementation may provide significant benefits for the operation of the whole company. There have now been a sufficient number of JIT implementations to demonstrate that JIT, when successfully implemented, will:

- reduce inventory levels, probably by about 50 per cent
- improve quality levels
- reduce scrap and rework rates
- reduce manufacturing lead times probably by 50–75 per cent
- improve customer service levels
- improve employee morale

These benefits are, of course, dependent on a successful implementation. Almost all JIT implementations will lead to some improvements, but the major benefits of JIT will probably only come about if the implementation is carried out in an informed and professional manner.

As indicated in chapter 3, JIT is not a software package. We do not purchase the computer software, input the correct data and collect an output to form the basis for managerial action. Software packages have formed the basis of several approaches to manufacturing management

Summary and Conclusion

that have been, and still are, prevalent. Examples are materials requirements planning (MRP), manufacturing resource planning (MRP II) and optimized production technology (OPT) [see Fox (1982)]. Although software may form part of the JIT implementation, it will not be the basis of it. Instead JIT is oriented towards improving the fundamental processes of manufacturing so as to improve the overall operation of the manufacturing enterprise.

JIT is not a strict methodology that can be defined as a series of equations or data relationships. Rather, it is a *philosophy* that leads to significant changes in the way that manufacturing management operates.

The philosophy of JIT is one that seeks continual improvement and for this to be effective the JIT philosophy must function at the core of company operations. Company personnel must therefore have assimilated the various aspects of the philosophy. This requires a significant change in attitudes and this book has stressed the major role that education plays, since without a good education programme the JIT implementation will at best be mediocre. This need for a change in company philosophy means that the JIT implementation cannot be achieved overnight but requires a good implementation plan, sufficient resources and a reasonable length of time (about one year for initial implementation).

We should expect that at the end of the initial implementation, provided that this is done well, manufacturing lead times, customer service and inventory levels will have improved. However, it is probably of far greater significance that, at the end of the initial implementations, the operation and attitudes of company personnel should have changed by absorbing the JIT philosophy. If this is not the case, the initial benefits of JIT will be lost as company personnel revert once more to their old habits.

Employee attitudes are a key measure of successful JIT implementation. Our goal is for *continual improvements* in operations for years to come which will only come about if the JIT philosophy is fully assimilated. The initial implementation should be considered only as a start with a successful implementation continuing to improve year after year.

The four principles of the JIT philosophy (chapter 3) are as follows:

1. *Attack fundamental problems.* JIT maintains there is little point in masking major problems such as capacity bottlenecks or poor quality vendors. It is far better to solve these fundamental problems and avoid a 'firefighting' style of management.

2. *Eliminate waste*. Waste is any activity that does not add value. Samples of such activities are inspection, transport and inventory. JIT stresses that these activities need to be eliminated to improve the overall operation of the company.
3. *Strive for simplicity*. Any approach that is adopted should be simple if it is to be effective. Previous approaches to manufacturing management have been based on complex management of a complex manufacturing system. By contrast, JIT implementation simplifies the flow of materials and then superimposes simple control.
4. *Devise systems to identify problems*. In order to solve fundamental problems, they need to be identified. A JIT implementation will include mechanisms that will bring problems to the fore. Examples of these mechanisms are statistical quality control (SQC), which monitors the manufacturing process and draws attention to any defect-producing trend, and pull/Kanban systems, which identify bottleneck production areas.

These four principles form the basis of any implementation but the way in which they are implemented may vary.

In this book five steps have been described which, when implemented in the proven path (Chapter 9), have been shown to be effective. These are:

Step 1: Getting the ball rolling. This step starts the whole implementation sequence and as such it sets the tone for the remainder of the implementation. The step can be broken down in to a number of stages: basic understanding, preliminary education, cost/benefit analysis, top management commitment, a go/no go decision, project team and pilot plant identification. The time scale involved in step 1 varies from company to company depending on the time taken to obtain top management commitment, the average being at least four months.

Step 2: Education – the make or break issue. A comprehensive education programme is essential to the success of a JIT implementation. JIT is concerned with a major shift in philosophy within the company and this can only materialize through JIT education.

Step 3: Process improvements. The manufacturing processes themselves have to be improved to produce small batches with short lead times. Process improvements include set-up time reduction and the move to product family flow lines.

Step 4: Improve the control. The simplified shop layouts typical of JIT implementations require simple but visible shop floor control to be fully effective. The mechanism generally favoured is the use of

Summary and Conclusion 129

pull/Kanban systems which are simple control arrangements that pull work through the factory rather than non-JIT shop floor control mechanisms which are primarily push. Advantages of these simple pull/Kanban systems are that their operation is visible to the shop floor personnel and that they automatically limit the amount of work-in-progress and hence the manufacturing lead times. These pull/Kanban systems will work most effectively where the shop floor culture is suitable and again this depends to a large extent on the education programme. In addition 'quality at source' needs to be pursued and the favoured method of achieving this is statistical process control.

Step 5: Vendor/customer links. This final step provides the enlargement necessary for JIT to encompass the entire system. Improving the vendor/customer links includes the gradual move to single-source high volume vendors. However (see Chapter 8), this should be done with extreme care to avoid the company becoming vulnerable.

To some extent several of these steps can be undertaken in parallel with a full implementation plan which the proven path promotes (Chapter 9). In this proven path, step 1 provides the essential preliminary phase. The time scale involved in step 1 varies widely from company to company as it involves obtaining top management commitment. Many companies manage step 1 in four months whilst others require several years. However, once step 1 is completed a company can then proceed to implement the other steps. The overall time scale for initial implementation is around twelve months from the end of step 1, and companies can expect to keep to this time scale.

The end result of implementing the five steps will be improvements in the manufacturing operation. However, what may be of greater long-term significance is that this initial implementation provides the basis of further improvements from year to year. It is the establishment of an infrastructure for on-going improvements that provides the major test of whether or not an implementation is successful.

Potential pitfalls

Major potential pitfalls in the implementation include:

1. *Failure to obtain top management commitment.* As difficult decisions have to be made during the JIT implementation, top management commitment is essential for effective implementation. Without it implementation can have disappointing results.
2. *Inadequate education programme.* Education is essential to change the

employee philosophy towards JIT. Fundamental changes are needed which require a comprehensive education programme to educate *everyone* associated with the implementation. This education programme should not only cover the initial implementation, but also continue over the following years to ensure reinforcement of the philosophy. A potential pitfall is to economize on the education programme and thereby jeopardize the entire JIT implementation.

3. *Inadequate external assistance.* The actual implementation of JIT will be achieved by company personnel, but a potential pitfall is to assume that they have all the knowledge and experience necessary. This is unlikely to be the case and the expertise of an external consultant will be invaluable in the implementation.

4. *Underestimating the task.* Implementing JIT is not an easy or soft option. It requires significant reorientation of company attitudes and some difficult and painful decisions often have to be made. JIT implementation is therefore a major task and this should be made clear to those involved at the outset.

5. *Implementation time scale too long or too short.* An implementation time scale that is too long risks enthusiasm being lost, whilst one that is too short risks false economies being made. A reasonable time scale for initial implementation is about one year after step 1 has been completed.

6. *Failing to integrate process and control improvements.* The improvements that can be made in both the processes and in control will not be substantial unless both are fully integrated. The operation of a JIT flow line relies on both process improvements (set-up time reduction, for example) and control improvements (pull/Kanban systems). If either of these happen in isolation the overall improvement will be small, if integrated the improvement can be large.

7. *Failure to write software.* Although JIT is not a software package, some software can help. Particular areas that may need software include loading the flow line and integration with the financial system.

8. *Rushing vendor links.* Sometimes a JIT implementation will emphasize changing vendors by reducing the number and by introducing sole-sourcing components. A potential pitfall is that the process is rushed, making the company heavily dependent on these vendors. Chapter 8 stressed that moving to sole-sourcing should happen slowly to test out the vendors.

9. *Failure to view JIT as an on-going process.* JIT improvements should

not be viewed as a short-term effort but should be expected to produce substantial returns for years to come. This will only happen if education, etc., is maintained.

The future

In companies that have aggressively followed the proven path JIT implementation will see many benefits. If we look ahead over the next ten years there will be an increasing number of companies moving to JIT implementation. Indeed, the cost/benefit ratio is so favourable that the implementation of JIT could be seen as perhaps inevitable in all manufacturing companies. The real question is not whether a particular company will implement JIT, but when.

For this to happen requires considerable changes in manufacturing management approaches. Managers with sound experience of implementing JIT will be in short supply and will be sought after. It is this shortage of expertise that will be the major constraint on the spread of JIT and it will probabaly be a major responsibility of universities, colleges and professional societies to provide support services to allow JIT to spread more quickly.

What about automation? The JIT philosophy stresses the improvements that can be obtained in work flow using simple (and usually inexpensive) approaches. The work flow needs to be improved first, and only when this has happened can automation be considered.

If automation takes place before the work flow has been improved the result is chaos. Automation in a JIT environment is therefore one of the last steps taken and even then it will involve very simple but flexible levels of automation. For example, simple pick-and-place robots will be used rather than more complex robotic equipment. It seems likely therefore that future manufacturing trends will incorporate this simple robotic equipment to aid the flow of work. The overall result will be an efficient, inexpensive but flexible manufacturing systems.

BIBLIOGRAPHY AND FURTHER READING

Abegglen, J C; How to Defend Your Business Against Japan, *Business Week* August 15, 1983.

Aggarwal S C: MRP, JIT, OPT, FMS? *Harvard Business Review* September–October, 1985.

Anderson J C, Schroeder R, Tupy S, White E; Material Requirements Planning Systems: The State of the Art, Fourt Quarter, 1982.

Buffa E S; *Meeting the Competitive Challenge* Dow Jones-Irwin, 1984.

Crosby P B; *Quality is Free; the Art of Making Quality Certain* McGraw-Hill, New York, 1979.

Ebrahimpour M, Schonberger R J; The Japanese Just-In-Time/Total Quality Control Production System: Potential For Developing Countries, *International Journal of Production Research* May–June, 1984.

Fox E; MRP, Kanban and OPT – What's Best? *APICS 25th Annual Conference Proceedings* 1982.

Garvin A; Quality on the Line, *Harvard Business Review* September–October, 1983.

Garwood Dave R; Explaining JIT, MRP II, Kanban, *P&IM Review and APICSNews* October 1984.

Hall W (CPIM); Stockless Production, *APICS* 1983.

Hays, R H, Wheelwright S C; *Restoring Our Competitive Edge – Competing Through Manufacturing* John Wiley, 1984.

Huang P Y, Rees L P, Taylor, III B W; *A Simulation Analysis of the Japanese Just-In-Time Technique (with Kanbans) for a Multiline, Multistage Production System, Decision Sciences* July 1983.

Johnson R W; Vendor Self-Inspection Sets the Stage for Just-In-Time Deliveries, *Quality Progress* November, 1984, pp. 46–47.

Jordan H; *The Challenge of Implementing A Zero Inventories Programme, P&IM Review and APICS News* June, 1985.

Klein J A; Why Supervisors Resist Employee Involvement, *Harvard Business Review* September-October, 1984, pp. 87–95.

McGuire J; Zero Inventory: Providing Focus For the Converging Manufacturing Strategies of the 80s, *Zero Inventory Philosophy & Practices Seminar Proceedings, APICS News.*

Monden Y; Adaptable Kanban System Helps Toyota Maintain Just-In-Time Production, *Industrial Engineering* May, 1981.

Monden Y; Toyota Production System *Industrial Engineering and Management Press* 1983.

Monden Y; A Simulation Analysis of the Japanese Just in Time Technique (with Kanbans) for a Multiline Multistage Production System: A Comment, *Decision Sciences* Summer 1984.

Monden Y; Applying Just In Time, *Industrial Engineering and Management Press* 1986.

Monden Y, Rinka, S; Innovations in Management, the Japanese Corporation, *Industrial Engineering and Management Press* 1985.

Orlicky J; *Material Requirements Planning* McGraw Hill, 1975.

Pendleton, W E (CPIM); Process Planning Optimization – The Key to ZI/JIT Success, *Zero Inventory Philosophy & Practices Seminar Proceedings, APICS* October, 1984.

Pipp F J; *Management commitment to quality: Xerox Corporation Quality Progress* August, 1983, pp. 12–17.

Sage L A; Just-In-Time: A Philosophy in Manufacturing Excellence, *Zero Inventory Philosophy & Practices Seminar Proceedings, APICS* October, 1984.

Schonberger R J; Japanese Manufacturing Techniques: nine hidden lessons in simplicity Free Press, New York 1979

Schonberger R J; Just in Time Production Systems: Replacing Complexity with Simplicity in Manufacturing Management, *Industrial Engineering* October, 1984.

Schonberger R J; *World Class Manufacturing* Free Press, New York, 1986.

Sepehri M; How Kanban is Used in an American Toyota Motor Facility, *Industrial Engineering* February, 1985.

Sepehri M; JIT and FMS – A working Team for Factory of the Future, P&IM Review and APICS News April, 1985.

Sepehri M, Quality Circles – A Vehicle for Just-in-Time Implementation, *Quality Progress* July, 1985.

Sepehri, M; *Just In Time; Not Just In Japan* Library of American Production, 1986.

Shingo S; *The Toyota Productions System* Japan Management Association, 1982.

Sugimori Y, Kusinoki K, Cho F, Uchikawa S; Toyota Production System and Kanban System – Materialization of Just-in-Time and Respect for Human System, *International Journal of Production Research* 1977, pp. 553–564.

Swoyer, Jr. (CPIM) OPT: Zero Inventory For All, *The Wall Street Journal* February 18, 1983.

Thomas P J; Waterman R H; *In Search of Excellence* Warner Books, 1982.

Vollman E, Berry L, Whybark D; *Manufacturing Planning and Control Systems* Dow Jones-Irwin, 1984.

Wallace T; *MRP II: Making it Happen* Oliver Wight Publications, 1985.

Wight O; *Manufacturing resource planning: MRP II* Oliver Wight Publications Van Nostrand Reinhold, 1981.

Index

American Production and Inventory Control Society (APICS), 23
Automation, 88, 131

Batch manufacturing, 7, 28, 42-3
  problems of, 17
Batch sizes, 35, 47
Bill of materials, 24
  accuracy of, 28
  inaccuracy of, 28-9
Birdcage layout, 86
Bottleneck machine or process, 37, 48-9, 94-5
Business planning, 26

Capacity requirements planning (CRP), 26
Commitment, 60, 129
Computers, 22, 26-8, 30, 33, 93
Continual improvement, 64-5, 69, 127
Control constraints, 92
Control improvement, 91-102, 128-30
Control mechanism, 91
Control of flow lines, 44
Control policies, 91
Control simplification, 92-3
Cost/benefit analysis, 53, 56-9
Cost savings, 103, 104
Customers. *See* Vendor/customer links

Data analysis, 33, 35
Decision making, 33, 34, 54-5
Deliveries
  rim system, 105-6
  spoke system, 105-6
Delivery performance, 106
Demand changes, 23
Demand forecasts, 20, 24, 30, 31
Dispatch list, 27

Economic Lot-Size Formula, 22
Economic order quantity (EOQ), 35, 47
Economic Order Quantity (EOQ) Formula, 22, 79
Economies of scale, 108
Education, 68-76, 128
  aims of, 70, 74
  content of, 73-5
  depth of coverage, 71
  design of programme, 73, 74, 129-30
  external courses, 71
  in-company courses, 71-2
  initial, 73, 74
  make or break issue, 68, 75
  necessity of, 70
  objective view, 74
  on-going, 73
  participants in, 71
  preliminary, 52-6
  requirements for, 69-70
  top management, 71
  workforce, 75
Employee attitudes, 127
Employee participation, 42, 69, 99

Flexible machines, 88
Flexible workforce, 85
Flow lines, 92, 94
  control of, 44
  process layout with, 78, 85-8
  product layout using, 44
  U-shaped, 86-7
Flow manufacture, 12-13
Freeze period, 29
Freight costs, 106

Go/no go checklist, 61
Go/no go decision, 53, 60-2

136 Index

Gradual improvement, 88
Group technology, 43

Harley Davidson, 29, 97, 104
Harris-Wilson Lot-Size Formula, 22
Human operations, 88

Implementation, 48, 60, 78, 128
  case study A, 119-22
  case study B, 122-4
  external assistance, 130
  five steps, 15-16, 114
  overall results of, 91
  potential pitfalls, 129
  proven path for, 114-25
  schedule, 64
  sequence of, 117-18
  time scale, 66-7, 115-16, 130
  underestimating task of, 130
Incremental efficiency increase, 48
Inspection, 38-9
Inventory control approach, 18-23
  disadvantages of, 22-3
  proactive, 23
  reactive, 23
Inventory costs, 39-40
Inventory holding costs, 23
Inventory levels, 17, 18, 21, 27, 31, 40, 45, 50, 51, 56, 57, 78, 83, 86, 106, 127
Inventory management, 106
Inventory monitoring and reordering, 18
Isolated island layout, 86

Jobbing manufacture, 13
Just-in-time (JIT)
  basic understanding, 53-4
  benefits of, 54, 58-9, 126
  capital expenditure, 49
  changes at shop floor level, 99
  comparison with traditional approaches, 33-5, 68-9, 92-3
  cost/benefit of, 49-50
  crucial feature of, 25
  essential objectives, 14-15
  freeze period, 29
  fundamental problems and solutions, 36-8
  future of, 131
  getting the ball rolling, 52-67
  implementing. *See* Implementation
  improvements resulting from, 122, 124
  methodology view, 35
  new direction of, 13
  overview, 32-51
  philosophy, 14, 35, 36-50, 73, 127
  principles, 127
  summary of, 50

Kanban system, 69, 70, 95-7
  dual card, 96
  single card, 97

Labour costs, 103
Lead times, 19-21, 25, 27, 29, 34, 38, 41, 43, 45, 48, 57, 95, 97, 99, 127
Long-term agreements, 108-9

Machine breakdowns, 83
Machine reliability, 36, 92
Manufacturing management, 11
  Japanese approach to, 12
  problems of, 12-13
  traditional approaches, 17-32
Manufacturing process
  improvements in, 77-90
  major categories, 12
Manufacturing resources planning (MRP II), 11, 17, 25-6, 127
  costs involved, 32
  effectiveness of, 27
  implementation of, 26-7, 30, 32, 49
  methodology shortcomings, 31
  problems of, 26-31
  re-implementation, 31
  software package, 33-4
Master product schedule (MPS), 24, 26
  accuracy of, 30
  effectiveness of, 31
  inaccuracy of, 29
  loss of confidence in, 30-1

Index

Material costs, 103
Material flow, simplicity requirement, 42-4
Materials requirements planning (MRP) 17, 23-5, 127
 effectiveness of, 27
 implementation of, 26-7, 30, 49
 linking with pull/Kanban systems, 97-9
 mechanisms of, 24
 methodology shortcomings, 31
 problems of, 26-31
 rationale of, 24
 re-implementation of, 31
 software package, 33-4
Monitoring, 64
Morale, 49
Multi-sourcing, 107-8

Obsolescence, 57
Obsolete stock, 23, 39
On-going process, 130
One-shot improvement, 64-5
Optimized production technology (OPT), 33-4, 127
 implementing, 49
Order quantities, 105
Ordering, 26

Pathfinders, 53, 54, 71
Performance ratings, 64
Personnel management, 60
Pilot plant identification, 65-6
Preventative maintenance, 45, 78, 82-5, 89
Problem identification, 49, 94, 128
Process improvements, 93, 128, 130
Process layout, 43
 with flow lines, 78, 85-8
Product changes, 28
Product layout with flow lines, 44
Production manager, 34
Production planning, 26
Productivity, 57
Productivity circles, 100
Project leader, 60, 62-3

Project team, 60, 74
 constitution of, 63
 selection of, 62-4
Pull/Kanban type systems, 44-7, 91, 93-5, 101, 117, 120, 128, 129
 advantages of, 91, 95
 linking MRP with, 97-9
 shop floor level, 99
Purchase order costs, 107
Purchasing, 104
 local *versus* distant suppliers, 110
Push/pull systems, 98

Quality control, 38-9, 91, 99-102
Quality improvement, 57

Reject rates, 34
Reorder level-reorder point-reorder quantities (ROL/ROP/ROQ), 33-4
Reorder point approach, 21
Reorder quantity formulae, 22
Rim system of deliveries, 105-6
River of inventory, 36, 51, 83
Robots, 131

Sales increases, 57
Saturn auto plant, 101
Set-up cost, 35
Set-up time, 47, 77
 reduction of, 78-82, 89, 92
Shojinka, 85, 86
Shop floor control, 99-101
Shop floor culture, 99-101
Shop floor expertise, 41
Short-term agreements, 108-9
Simplicity requirement, 42-9, 128
 material flow, 42-4
Single-sourcing, 107-8
Software packages, 33, 127, 130
Spoke system of deliveries, 105-6
Statistical quality control, 47, 91, 102, 128
Stock-piles, 20
Stockturn objectives, 37
Strategic area, 33-4

Streamlined bureaucracy, 106
Suggestion schemes, 41, 100
Suppliers, local *versus* distant, 110

Tactical area, 34
Time scale for implementation, 66-7, 115-16, 130
Tolerances, approaches to, 39
Top management
    commitment of, 60, 129
    education, 71
Total preventative maintenance (TPM), 83-5, 88
Toyota, 46, 47, 48, 80, 95, 96, 97, 101
Training programme, 70

U-shaped flow lines, 86-7

Vendor/customer links, 103-13, 130
    customer service, 127
    features to be considered in, 104-7
    implementation, 110-12
    local *versus* distant suppliers, 110
    multi-sourcing *versus* single-sourcing, 107
    short- *versus* long-term agreements, 108-9
Vendor performance, 106
Video tapes, 82

Waste elimination, 38-42, 105, 128
Work flow improvement, 131
Work-in-progress, 31, 34, 43, 45, 48, 49, 56, 57, 83, 86, 95, 97, 99